PSSA
Math Practice
Grade 4

Complete Content Review Plus
2 Full-length PSSA Math Tests

Elise Baniam - Michael Smith

PSSA Math Practice Grade 4

Published in the United State of America By

The Math Notion

Email: info@Mathnotion.com

Web: WWW.MathNotion.com

Copyright © 2020 by the Math Notion. All rights reserved. No part of this publication may be reproduced, stored in a retrieval system, or transmitted in any form or by any means, electronic, mechanical, photocopying, recording, scanning, or otherwise, except as permitted under Section 107 or 108 of the 1976 United States Copyright Ac, without permission of the author.

All inquiries should be addressed to the Math Notion.

ISBN: 978-1-63620-014-9

About the Author

Elise Baniam has been a math instructor for over a decade now. She graduated in Mathematics. Since 2006, Elise has devoted his time to both teaching and developing exceptional math learning materials. As a Math instructor and test prep expert, Elise has worked with thousands of students. She has used the feedback of her students to develop a unique study program that can be used by students to drastically improve their math score fast and effectively.

- **– SAT Math Workbook**
- **– ACT Math Workbook**
- **– ISEE Math Workbooks**
- **– SSAT Math Workbooks**
- **–many Math Education Workbooks**
- **– and some Mathematics books …**

As an experienced Math teacher, Mrs. Baniam employs a variety of formats to help students achieve their goals: she teaches students in large groups, and she provides training materials and textbooks through her website and through Amazon.

You can contact Elise via email at:
Elise@Mathnotion.com

PSSA Math Practice Grade 4

Get the Targeted Practice You Need to Excel on the Math Section of the PSSA Test Grade 4!

PSSA Math Practice Book Grade 4 is **an excellent investment in your future** and the best solution for students who want to maximize their score and minimize study time. Practice is an essential part of preparing for a test and improving a test taker's chance of success. The best way to practice taking a test is by going through lots of PSSA math questions.

High-quality mathematics instruction ensures that students become problem solvers. We believe all students can develop deep conceptual understanding and procedural fluency in mathematics. In doing so, through this math workbook we help our students grapple with real problems, think mathematically, and create solutions.

PSSA Math Practice Book allows you to:

- Reinforce your strengths and improve your weaknesses
- Practice **2500+ realistic** PSSA math practice questions
- Exercise math problems in a variety of formats that provide intensive practice
- Review and study **Two Full-length PSSA Practice Tests** with detailed explanations

...and much more!

This Comprehensive PSSA Math Practice Book is carefully designed to provide only that **clear and concise information** you need.

WWW.MathNotion.com

… So Much More Online!

✓ FREE Math Lessons

✓ More Math Learning Books!

✓ Mathematics Worksheets

✓ Online Math Tutors

For a PDF Version of This Book

Please Visit WWW.MathNotion.com

Contents

Chapter 1: Place Value and Number Sense 11
    Numbers in Standard Form 12
    Number in Expand Form 13
    Odd or Even 14
    Compare Whole Numbers 15
    Pattern 16
    Round Whole Numbers 17
    Answer key Chapter 1 18

Chapter 2: Adding and Subtracting 21
    Adding 3–Digit Numbers 22
    Adding 4–Digit Numbers 23
    Estimate Sums 24
    Subtracting 3–Digit Numbers 25
    Subtracting 4–Digit Numbers 26
    Estimate Differences 27
    Subtract from Whole Thousands. 28
    Answer key Chapter 2 29

Chapter 3: Multiplication and Division 31
    Multiplication Whole Numbers 32
    Multiply Tens and Hundreds. 33
    Estimate Products 34
    Multiplication Missing Numbers 35
    Long Division by One Digit 36
    Division with Remainders 37
    Dividing Tens and Hundreds 38
    Division Missing Number 39
    Answer key Chapter 3 40

Chapter 4: Number Theory 43
    Factoring 44
    Prime Factorization 45
    Divisibility Rule 46

Great Common Factor (GCF) ... 47
Least Common Multiple (LCM) ... 48
Distributive Property ... 49
Answer key Chapter 4 ... 50

Chapter 5: Patterns ... 53
Repeating Pattern ... 54
Growing Patterns ... 55
Patterns: Numbers ... 56
Find a Rule ... 57
Algebraic Thinking ... 58
Answers of Worksheets – Chapter 5 ... 59

Chapter 6: Fractions and Mix Numbers ... 61
Adding Fractions – Like Denominator ... 62
Adding Fractions – Unlike Denominator ... 63
Subtracting Fractions – Like Denominator ... 64
Subtracting Fractions – Unlike Denominator ... 65
Converting Mix Numbers ... 66
Converting improper Fractions ... 67
Adding Mix Numbers ... 68
Subtracting Mix Numbers ... 69
Simplify Fractions ... 70
Multiplying Fractions ... 71
Multiplying Mixed Number ... 72
Comparing Fractions ... 73
Answer key Chapter 6 ... 74

Chapter 7: Decimal ... 79
Graph Decimals ... 80
Round Decimals ... 81
Decimals Addition ... 82
Decimals Subtraction ... 83
Decimals Multiplication ... 84
Decimal Division ... 85
Comparing Decimals ... 86
Convert Fraction to Decimal ... 87

Answer key Chapter 7 88

Chapter 8: Measurement 91

Reference Measurement 92
Metric Length Measurement 93
Customary Length Measurement 93
Metric Capacity Measurement 94
Customary Capacity Measurement 94
Metric Weight and Mass Measurement 95
Customary Weight and Mass Measurement 95
Time 96
Answers of Worksheets – Chapter 8 97

Chapter 9: Symmetry and Transformations 99

Line Segments 100
Parallel, Perpendicular and Intersecting Lines 101
Identify Lines of Symmetry 102
Lines of Symmetry 103
Identify Three–Dimensional Figures 104
Vertices, Edges, and Faces 105
Identify Faces of Three–Dimensional Figures 106
Answers of Worksheets – Chapter 9 107

Chapter 10: Geometry 111

Identifying Angles 112
Polygon Names 113
Triangles 114
Quadrilaterals and Rectangles 115
Area and Perimeter of Square 116
Area and Perimeter of Rectangle 117
Area and Perimeter of Triangle 118
Perimeter of Polygon 119
Answer key Chapter 10 120

Chapter 11: Data and Graphs 121

Tally and Pictographs 122
Stem–And–Leaf Plot 123
Dot plots 124

Coordinate Plane 125
Bar Graph 126
Line Graphs 127
Answer key Chapter 11 128

PSSA Test Review 131

PSSA Practice Test 1 135
PSSA Practice Test 2 145

Answers and Explanations 153

Answer Key 155
Practice Test 1 157
Practice Test 2 160

Chapter 1:
Place Value and Number Sense

Numbers in Standard Form

Write the number in standard form.

1) 10 million 208 thousand 24

   _____

2) 72 million 9 thousand 708

   _____

3) 121 million 24 thousand 453

   _____

4) 541 million 75 thousand 127

   _____

5) 90 billion 15 million 68 thousand 15

   _____

6) 12 billion 120 million 5

   _____

7) 8 billion 114 million 88 thousand

   _____

8) 16 billion 28 thousand 785

   _____

9) 75 billion 159 thousand 324

   _____

10) 41 billion 3 million 8 thousand 25

    _____

11) 16 billion 129 thousand 989

    _____

12) 65 billion 220 million 6 thousand 2

    _____

13) 785 million 124 thousand 97

    _____

14) 33 billion 104 million 11 thousand 57

    _____

15) 95 billion 424 million

    _____

16) 27 billion 77 million 9 thousand 150

    _____

Number in Expand Form

Write the number in expand form.

1) 956: _____.

2) 3,800: _____.

3) 52,457: _____.

4) 60,070: _____.

5) 409,389: _____.

6) 76,805: _____.

7) 745,321: _____.

8) 8,146: _____.

9) 19,037: _____.

10) 52,799: _____.

11) 5,125: _____.

12) 400,544: _____.

13) 600,700: _____.

14) 3,080,000: _____.

Odd or Even

Write odd or even.

1) 19 _____

2) 81 _____

3) 456 _____

4) 852 _____

5) 953 _____

6) 183 _____

7) 987 _____

8) 540 _____

9) 777 _____

10) 544 _____

11) 33 _____

12) 4,458 _____

13) 15,159 _____

14) 9,357 _____

15) 3,000 _____

16) 14 _____

17) 257 _____

18) 660 _____

19) 45,789 _____

20) 15,300 _____

21) 452 _____

22) 49,459 _____

23) 84 _____

24) 7,700 _____

25) 6,451 _____

26) 985 _____

Compare Whole Numbers

Compare, writing <, >, or = between the numbers.

1) 40,420 ☐ 41,004

2) 29,460 ☐ 29,640

3) 78,920 ☐ 87,290

4) 34,570 ☐ 33,750

5) 96,328 ☐ 96,238

6) 85,843 ☐ 85,840

7) 76,584 ☐ 76,854

8) 72,998 ☐ 72,989

9) 37,467 ☐ 37,567

10) 48,878 ☐ 49,878

11) 56,660 ☐ 65,660

12) 73,898 ☐ 69,899

13) 89,990 ☐ 98,110

14) 84,760 ☐ 84,670

15) 26,680 ☐ 26,860

16) 86,440 ☐ 86,440

17) 158,980 ☐ 158,890

18) 201,807 ☐ 201,807

19) 243,240 ☐ 243,420

20) 345,566 ☐ 354,655

21) 187,158 ☐ 196,001

22) 137,983 ☐ 137,895

23) 278,788 ☐ 249,988

24) 194,854 ☐ 194,845

25) 219,390 ☐ 291,110

26) 305,288 ☐ 299,999

27) 317,857 ☐ 371,857

28) 405,710 ☐ 405,170

Pattern

Continue this pattern for four more numbers:

1) 1,400; 1,250; 1,100; 950; _____

2) 3,700; 3,500; 3,300; 3,100; _____

3) 4,200; 3,850; 3,500; 3,150; _____

4) 1,650; 1,530; 1,410; 1,290; _____

5) 2,900; 2,650; 2,400; 2,150; _____

6) 5,000; 4,600; 4,200; 3,800; _____

7) 3,950; 3,700; 3,450; 3,200; _____

8) 2,800; 2,675; 2,550; 2,425; _____

9) 3,850; 3,550; 3,250; 2,950; _____

10) 4,700; 4,500; 4,300; 4,100; _____

11) Write a list of five numbers that follows this pattern: Start at 250 and add 200 each time.

_____

Round Whole Numbers

Round to the place of the underlined digit.

1) 7,46<u>7</u>,589 ≈ _____

2) 54<u>6</u>,125 ≈ _____

3) 9,18<u>7</u>,208 ≈ _____

4) 15,68<u>5</u>,807 ≈ _____

5) 5,45<u>4</u>,676 ≈ _____

6) 3,58<u>8</u>,975 ≈ _____

7) 8,368,<u>5</u>19 ≈ _____

8) 27,754,<u>7</u>69 ≈ _____

9) 42,654,<u>4</u>11 ≈ _____

10) 7, <u>6</u>21,879 ≈ _____

11) 19,78<u>8</u>,987 ≈ _____

12) 4,28<u>6</u>,850 ≈ _____

13) 9,2<u>7</u>3,778 ≈ _____

14) 6,48<u>4</u>,684 ≈ _____

15) 5,157,<u>6</u>28 ≈ _____

16) 8,66<u>7</u>,885 ≈ _____

17) 3,56<u>7</u>,980 ≈ _____

18) 8,369,4<u>3</u>2 ≈ _____

19) 24,25<u>6</u>,880 ≈ _____

20) 5,2<u>2</u>9,758 ≈ _____

21) 6,9<u>8</u>7,422 ≈ _____

22) 4,87<u>7</u>,391 ≈ _____

Answer key Chapter 1

Numbers in Standard Form

1) 10,208,024
2) 72,009,708
3) 121,024,453
4) 541,075,127
9) 75,000,159,324
10) 41,003,008,025
11) 16,000,129,989
12) 65,220,006,002
5) 90,015,068,015
6) 12,120,000,005
7) 8,114,088,000
8) 16,000,028,785
13) 785,124,097
14) 33,104,011,057
15) 95,424,000,000
16) 27,077,009,150

Numbers in Expand Form

1) $(9 \times 100) + (5 \times 10) + 6$
2) $(3 \times 1{,}000) + (8 \times 100)$
3) $(5 \times 10{,}000) + (2 \times 1{,}000) + (4 \times 100) + (5 \times 10) + 7$
4) $(6 \times 10{,}000) + (7 \times 10) + 0$
5) $(4 \times 100{,}000) + (9 \times 1{,}000) + (3 \times 100) + (8 \times 10) + 9$
6) $(7 \times 10{,}000) + (6 \times 1{,}000) + (8 \times 100) + 5$
7) $(7 \times 100{,}000) + (4 \times 10{,}000) + (5 \times 1{,}000) + (3 \times 100) + (2 \times 10) + 1$
8) $(8 \times 1{,}000) + (1 \times 100) + (4 \times 10) + 6$
9) $(1 \times 10{,}000) + (9 \times 1{,}000) + (3 \times 10) + 7$
10) $(5 \times 10{,}000) + (2 \times 1{,}000) + (7 \times 100) + (9 \times 10) + 9$
11) $(5 \times 1{,}000) + (1 \times 100) + (2 \times 10) + 5$
12) $(4 \times 100{,}000) + (5 \times 100) + (4 \times 10) + 4$
13) $(6 \times 100{,}000) + (7 \times 100)$
14) $(3 \times 1{,}000{,}000) + (8 \times 10{,}000)$

Odd or Even

1) Odd
2) Odd
3) Even
4) Even
5) Odd
6) Odd
7) Odd
8) Even
9) Odd
10) Even
11) Odd
12) Even

PSSA Math Practice Grade 4

13) Odd
14) Odd
15) Even
16) Even
17) Odd
18) Even
19) Odd
20) Even
21) Even
22) Odd
23) Even
24) Even
25) Odd
26) Odd

Compare Whole Numbers

1) <
2) <
3) <
4) >
5) >
6) >
7) <
8) >
9) <
10) <
11) <
12) >
13) <
14) >
15) <
16) =
17) >
18) =
19) <
20) <
21) <
22) >
23) >
24) >
25) <
26) >
27) <
28) >

Pattern

1) 800; 650; 500; 350
2) 2,900; 2,700; 2,500; 2,300
3) 2,800; 2,450; 2,100; 1,750
4) 1,170; 1,050; 930; 810
5) 1,900; 1,650; 1,400; 1,150
6) 3,400; 3,000; 2,600; 2,200
7) 2,950; 2,700; 2,450; 2,200
8) 2,300; 2,175; 2,050; 1,925
9) 2,650; 2,350; 2,050; 1,750
10) 3,900; 3,700; 3,500; 3,300
11) 250; 450; 650; 850; 1,050

Round whole number

1) 7,470,000
2) 546,000
3) 9,187,000
4) 15,686,000
5) 5,455,000
6) 3,589,000
7) 8,368,500
8) 27,754,800
9) 42,654,400
10) 7,600,000
11) 19,789,000
12) 4,287,000
13) 9,270,000
14) 6,485,000
15) 5,157,600
16) 8,668,000
17) 3,568,000
18) 8,369,430
19) 24,257,000
20) 5,230,000
21) 6,990,000
22) 4,877,000

Chapter 2: Adding and Subtracting

Adding 3–Digit Numbers

Find each sum.

1) 526 + 236

2) 725 + 130

3) 425 + 153

4) 563 + 125

5) 453 + 230

6) 398 + 120

7) 689 + 456

8) 863 + 325

9) 965 + 865

10) 369 + 120

11) 187 + 125

12) 389 + 150

13) 469 + 156

14) 360 + 150

15) 689 + 263

16) 890 + 345

17) 720 + 215

18) 680 + 230

Adding 4–Digit Numbers

Add.

1) 2,135
   + 5,236
   _____

2) 4,369
   + 1,356
   _____

3) 6,598
   + 2,325
   _____

4) 3,125
   +4,035
   _____

5) 4,135
   +2,194
   _____

6) 5,036
   +2,365
   _____

7) 3,236
   +2,369
   _____

8) 6,320
   +3,765
   _____

9) 3,890
   +3,567
   _____

Find the missing numbers.

10) 1,155 + __ = 1,469

11) 400 + 3,000 = __

12) 5,200 + __ = 7,300

13) 555 + __ = 1,886

14) __ + 920 = 1,550

15) __ + 2,670 = 4,230

16) 689,505 = 80,000 + 600,000 + 5 + _____ + 9,000

17) 750,678 = 50,000 + 700,000 + 8 + _____ + 600

18) 574,962 = 70,000 + 500,000 + 2 + _____ + 900 + 60

PSSA Math Practice Grade 4

Estimate Sums

Estimate the sum by rounding each added to the nearest ten.

1) 36 + 9 =

2) 29 + 46 =

3) 36 + 12 =

4) 37 + 38 =

5) 12 + 35 =

6) 38 + 13 =

7) 48 + 25 =

8) 36 + 77 =

9) 45 + 86 =

10) 62 + 58 =

11) 45 + 36 =

12) 52 + 18 =

13) 35 + 59 =

14) 38 + 65 =

15) 87 + 82 =

16) 18 + 69 =

17) 65 + 64 =

18) 33 + 26 =

19) 73 + 48 =

20) 35 + 64 =

21) 13 + 93 =

22) 63 + 52 =

23) 164 + 142 =

24) 54 + 77 =

Subtracting 3–Digit Numbers

Find the difference.

1) 756 − 236

2) 693 − 130

3) 425 − 153

4) 365 − 125

5) 493 − 230

6) 398 − 120

7) 989 − 756

8) 863 − 325

9) 965 − 465

10) 369 − 120

11) 159 − 125

12) 789 − 450

13) 469 − 156

14) 960 − 250

15) 689 − 358

16) 890 − 345

17) 929 − 115

18) 999 − 130

Subtracting 4–Digit Numbers

Subtract.

1) 3,130 − 1,134

2) 3,356 − 2,870

3) 5,986 − 2,678

4) 6,987 − 6,422

5) 5,362 − 3,331

6) 7,365 − 2,212

7) 8,356 − 5,712

8) 8,350 − 2,729

9) 6,117 − 1,216

Find the missing number.

10) 4,223 − __ = 2,320

11) 5,856 − __ = 4,245

12) 1,136 − 689 = __

13) 4,200 − __ = 2,450

14) 5,870 − 2,650 = __

15) 6,360 − 4,320 = __

16) 8,165 − _____ = 4,303

17) 5,060 − 1,867 = __

18) Bob had $3,486 invested in the stock market until he lost $2,198 on those investments. How much money does he have in the stock market now?

Estimate Differences

Estimate the difference by rounding each number to the nearest ten.

1) $58 - 23 =$

2) $34 - 24 =$

3) $75 - 48 =$

4) $43 - 24 =$

5) $69 - 46 =$

6) $42 - 23 =$

7) $77 - 47 =$

8) $49 - 28 =$

9) $94 - 48 =$

10) $79 - 59 =$

11) $68 - 26 =$

12) $83 - 37 =$

13) $73 - 43 =$

14) $58 - 42 =$

15) $82 - 52 =$

16) $65 - 43 =$

17) $99 - 81 =$

18) $42 - 24 =$

19) $58 - 47 =$

20) $89 - 28 =$

21) $81 - 65 =$

22) $68 - 14 =$

23) $76 - 6 =$

24) $78 - 31 =$

Subtract from Whole Thousands.

Find the difference.

1) $3{,}000 - 10 =$ ___

2) $4{,}000 - 5 =$ ___

3) $2{,}000 - 8 =$ ___

4) $5{,}000 - 30 =$ ___

5) $7{,}000 - 7 =$ ___

6) $6{,}000 - 15 =$ ___

7) $8{,}000 - 40 =$ ___

8) $9{,}000 - 5 =$ ___

9) $2{,}000 - 8 =$ ___

10) $5{,}000 - 30 =$ ___

11) $7{,}000 - 200 =$ ___

12) $6{,}000 - 2 =$ ___

13) $4{,}000 - 20 =$ ___

14) $8{,}000 - 200 =$ ___

15) $5{,}000 - 100 =$ ___

16) $6{,}000 - 80 =$ ___

17) $5{,}000 - 70 =$ ___

18) $7{,}000 - 200 =$ ___

19) $9{,}000 - 300 =$ ___

20) $2{,}000 - 8 =$ ___

21) $4{,}000 - 10 =$ ___

22) $8{,}000 - 50 =$ ___

23) $3{,}000 - 90 =$ ___

24) $1{,}000 - 6 =$ ___

25) $5{,}000 - 5 =$ ___

26) $8{,}000 - 90 =$ ___

27) $9{,}000 - 30 =$ ___

28) $2{,}000 - 60 =$ ___

Answer key Chapter 2

Adding three–digit numbers

1) 762
2) 855
3) 578
4) 688
5) 683
6) 518
7) 1,145
8) 1,188
9) 1,830
10) 489
11) 312
12) 539
13) 625
14) 510
15) 952
16) 1,235
17) 935
18) 910

Adding 4–digit numbers

1) 7,371
2) 5,725
3) 8,923
4) 7,160
5) 6,329
6) 7,401
7) 5,605
8) 10,085
9) 7,457
10) 314
11) 3,400
12) 2,100
13) 1,331
14) 630
15) 1,560
16) 500
17) 70
18) 4,000

Estimate sums

1) 50
2) 80
3) 50
4) 80
5) 50
6) 50
7) 80
8) 120
9) 140
10) 120
11) 90
12) 70
13) 100
14) 110
15) 170
16) 90
17) 130
18) 60
19) 120
20) 100
21) 100
22) 110
23) 300
24) 130

Subtracting 3–digit numbers

1) 520
2) 563
3) 272
4) 240
5) 263
6) 278
7) 233
8) 538
9) 500
10) 249
11) 34
12) 339
13) 313
14) 710
15) 331
16) 545
17) 814
18) 869

Subtracting 4–digit numbers

1) 1,996
2) 486
3) 3,308

4) 565
5) 2,031
6) 5,153
7) 2,644
8) 5,621
9) 4,901
10) 1,903
11) 1,611
12) 447
13) 1,750
14) 3,220
15) 2,040
16) 3,862
17) 3,193
18) 1,288

Estimate differences

1) 40
2) 10
3) 30
4) 20
5) 20
6) 20
7) 30
8) 20
9) 40
10) 20
11) 40
12) 40
13) 30
14) 20
15) 30
16) 30
17) 20
18) 20
19) 10
20) 60
21) 10
22) 60
23) 70
24) 50

Subtract from Whole Thousands

1) 2,990
2) 3,995
3) 1,992
4) 4,970
5) 6,993
6) 5,985
7) 7,960
8) 8,995
9) 1,992
10) 4,970
11) 6,800
12) 5,998
13) 3,980
14) 7,800
15) 4,900
16) 5,920
17) 5,930
18) 6,800
19) 8,700
20) 1,992
21) 3,990
22) 7,950
23) 2,910
24) 994
25) 4,995
26) 7,910
27) 8,970
28) 1,940

Chapter 3: Multiplication and Division

Multiplication Whole Numbers

Find the answers.

1) 53 × 12

2) 46 × 10

3) 17 × 12

4) 45 × 14

5) 48 × 12

6) 45 × 21

7) 12 × 13

8) 42 × 20

9) 140 × 7

10) 564 × 4

11) 363 × 4

12) 36 × 20

13) 345 × 23

14) 725 × 30

15) 364 × 25

16) Emily has 17 candy bars. She divided each bar into 7 equal pieces to share with her colleagues. How many colleagues does Emily have? _____

17) Harper packaged cupcake in boxes of 12. She filled 36 boxes. How many cupcakes does Harper have? _____

Multiply Tens and Hundreds.

Multiply, and find the missing factors.

1) 50 × 6 = _____

2) 7 × 400 = _____

3) 30 × 9 = _____

4) 80 × 70 = _____

5) 6 × 400 = _____

6) 70 × 90 = _____

7) 20 × 600 = _____

8) 10 × 900 = _____

9) 60 × 800 = _____

10) 7 × 700 = _____

11) _____ × 4 = 320

12) _____ × 8 = 6,400

13) _____ × 7 = 2,100

14) _____ × 9 = 720

15) _____ × 3 = 3,600

16) _____ × 600 = 5,400

17) _____ × 70 = 2,800

18) _____ × 30 = 2,700

19) _____ × 200 = 1,400

20) _____ × 300 = 12,000

21) 90 × _____ = 4,500

22) 40 × _____ = 2,400

23) 80 × _____ = 8,000

24) 60 × _____ = 420

25) 30 × _____ = 15,000

26) 700 × _____ = 6,3000

27) 50 × _____ = 3,000

28) 300 × _____ = 18,000

Estimate Products

Estimate the products.

1) 38 × 17 =

2) 13 × 16 =

3) 23 × 16 =

4) 23 × 12 =

5) 65 × 21 =

6) 38 × 71 =

7) 42 × 92 =

8) 15 × 39 =

9) 23 × 14 =

10) 73 × 33 =

11) 43 × 24 =

12) 49 × 13 =

13) 58 × 33 =

14) 82 × 56 =

15) 52 × 77 =

16) 26 × 58 =

17) 34 × 38 =

18) 33 × 47 =

19) 32 × 36 =

20) 35 × 47 =

21) 75 × 53 =

22) 29 × 11 =

23) 53 × 11 =

24) 94 × 36 =

Multiplication Missing Numbers

Find the missing numbers.

1) 20 × __ = 80

2) 16 × __ = 48

3) __ × 12 = 96

4) 12 × __ = 48

5) __ × 17 = 102

6) 15 × __ = 135

7) 6 × __ = 48

8) 80 × __ = 2,400

9) 12 × 7 = __

10) 36 × 5 = __

11) 22 × 4 = __

12) 69 × 3 = __

13) __ × 45 = 270

14) 9 × __ = 360

15) 70 × __ = 280

16) 32 × __ = 256

17) __ × 30 = 270

18) 25 × 5 = __

19) __ × 13 = 169

20) 19 × __ = 228

21) 40 × 6 = __

22) 50 × 3 = __

23) __ × 26 = 832

24) 18 × __ = 216

25) Emily has 24 candy bars. She divided each bar into 7 equal pieces to share with her colleagues. How many colleagues does Emily have? _____

Long Division by One Digit

Find the quotient.

1) $5\overline{)100}=$

2) $8\overline{)64}=$

3) $13\overline{)169}=$

4) $3\overline{)24}=$

5) $12\overline{)144}=$

6) $8\overline{)48}=$

7) $2\overline{)12}=$

8) $7\overline{)21}=$

9) $9\overline{)468}=$

10) $5\overline{)30}=$

11) $4\overline{)36}=$

12) $13\overline{)65}=$

13) $8\overline{)56}=$

14) $9\overline{)90}=$

15) $8\overline{)112}=$

16) $24\overline{)360}=$

17) $2\overline{)36}=$

18) $8\overline{)24}=$

19) $4\overline{)60}=$

20) $9\overline{)153}=$

21) $6\overline{)114}=$

22) $5\overline{)90}=$

23) $10\overline{)1,170}=$

24) $11\overline{)462}=$

25) $4\overline{)540}=$

26) $8\overline{)640}=$

27) $8\overline{)216}=$

28) $8\overline{)112}=$

29) $15\overline{)495}=$

30) $20\overline{)400}=$

31) $11\overline{)484}=$

32) $10\overline{)800}=$

33) $2\overline{)64}=$

34) $3\overline{)48}=$

35) $4\overline{)76}=$

36) $12\overline{)720}=$

37) $8\overline{)1,160}=$

38) $6\overline{)750}=$

39) $9\overline{)3,168}=$

40) $4\overline{)812}=$

41) $5\overline{)1,025}=$

42) $3\overline{)2,589}=$

Division with Remainders

Find the quotient with remainder.

1) 6)38

2) 5)29

3) 8)67

4) 3)10

5) 7)53

6) 3)17

7) 12)146

8) 21)444

9) 4)22

10) 5)48

11) 10)71

12) 8)24

13) 7)51

14) 9)84

15) 6)40

16) 12)126

17) 15)228

18) 11)177

19) 13)38

20) 3)20

21) 13)222

22) 5)46

23) 4)9

24) 9)1450

25) 96)194

26) 36)183

27) 38)230

28) 146)443

29) 42)1,766

30) 92)554

31) 210)632

32) 135)810

33) 6)79

34) 13)161

35) 126)885

36) 85)853

37) 125)1502

38) 11)4,832

39) 8)2,691

40) 7)953

41) 3)2,265

42) 4)6,744

Dividing Tens and Hundreds

Find answers.

1) $2000 \div 200$

2) $1600 \div 20$

3) $900 \div 100$

4) $3,200 \div 800$

5) $4,800 \div 800$

6) $900 \div 300$

7) $2,400 \div 800$

8) $4,500 \div 900$

9) $6,800 \div 200$

10) $10,000 \div 200$

11) $8,100 \div 300$

12) $8,000 \div 500$

13) $1,200 \div 200$

14) $6,600 \div 600$

15) $7,200 \div 600$

16) $1,800 \div 200$

17) $27,000 \div 900$

18) $9,900 \div 300$

19) $7,200 \div 100$

20) $9,000 \div 120$

21) $9,000 \div 3,000$

22) $16,000 \div 40$

23) $210 \div 30$

24) $560 \div 70$

Division Missing Number

Find each missing number.

1) 16 ÷ __ = 2

2) __ ÷ 8 = 6

3) 18 ÷ __ = 2

4) __ ÷ 5 = 9

5) 32 ÷ __ = 4

6) __ ÷ 9 = 8

7) 40 ÷ __ = 5

8) 240 ÷ 16 = __

9) 99 ÷ __ = 9

10) 80 ÷ 10 = __

11) 24 ÷ __ = 3

12) 42 ÷ __ = 6

13) __ ÷ 8 = 7

14) 120 ÷ 40 = __

15) 18 ÷ __ = 1

16) 60 ÷ __ = 6

17) __ ÷ 14 = 9

18) __ ÷ 11 = 13

19) 70 ÷ __ = 7

20) __ ÷ 10 = 3

21) 49 ÷ 7 = __

22) 100 ÷ 10 = __

23) 14 ÷ 14 = __

24) 625 ÷ __ = 25

25) Linda planted 180 seeds. She wants to put them in equal numbers on 6 rows. How many seed can she put on a row? _____ seeds

26) If dividend is 144 and the quotient is 16, then what is the divisor? _____

Answer key Chapter 3

Multiplication whole numbers

1) 636	7) 156	13) 7,935
2) 460	8) 840	14) 21,750
3) 204	9) 980	15) 9,100
4) 630	10) 2,256	16) 119
5) 576	11) 1,452	17) 432
6) 945	12) 720	

Multiply Tens and Hundreds300

1) 2,800	10) 80	19) 40
2) 270	11) 800	20) 50
3) 560	12) 300	21) 60
4) 2,400	13) 80	22) 100
5) 6,300	14) 12,00	23) 7
6) 12,000	15) 9	24) 500
7) 9,000	16) 40	25) 90
8) 48,000	17) 90	26) 60
9) 4,900	18) 7	27) 60

Estimate products

1) 800	9) 200	17) 1,200
2) 200	10) 2,100	18) 1,500
3) 400	11) 800	19) 1,200
4) 200	12) 500	20) 2,000
5) 1,400	13) 1,800	21) 4,000
6) 2,800	14) 4,800	22) 3,000
7) 3,600	15) 4,000	23) 500
8) 800	16) 1,800	24) 3,600

Multiplication Missing Numbers

1) 4	4) 4	7) 8
2) 3	5) 6	8) 30
3) 8	6) 9	9) 84

WWW.MathNotion.com

10) 180
11) 88
12) 207
13) 6
14) 40
15) 4

16) 8
17) 9
18) 125
19) 13
20) 12
21) 240

22) 150
23) 32
24) 12
25) 168

Long Division by One Digit

1) 20
2) 8
3) 13
4) 8
5) 12
6) 6
7) 6
8) 3
9) 52
10) 6
11) 9
12) 5
13) 7
14) 10

15) 14
16) 15
17) 18
18) 3
19) 15
20) 17
21) 19
22) 18
23) 117
24) 42
25) 135
26) 80
27) 27
28) 14

29) 33
30) 20
31) 44
32) 80
33) 32
34) 16
35) 19
36) 60
37) 145
38) 125
39) 352
40) 203
41) 205
42) 863

Division with Remainders

1) 6 R2
2) 5 R4
3) 8 R3
4) 3 R1
5) 7 R4
6) 5 R2
7) 12 R2
8) 21 R3
9) 4 R2

10) 9 R3
11) 7 R1
12) 3 R0
13) 7 R2
14) 9 R3
15) 6 R4
16) 10 R6
17) 15 R3
18) 16 R1

19) 2 R12
20) 6 R2
21) 17 R1
22) 9 R1
23) 2 R1
24) 161 R1
25) 2 R2
26) 5 R3
27) 6 R2

PSSA Math Practice Grade 4

28) 3 R5
29) 42 R2
30) 6 R2
31) 3 R2
32) 6 R0
33) 13 R1
34) 12 R5
35) 7 R3
36) 10 R3
37) 12 R2
38) 439 R3
39) 336 R3
40) R8
41) 753 R6
42) 1,685 R4

Dividing Tens and Hundreds

1) 10
2) 80
3) 9
4) 4
5) 6
6) 3
7) 3
8) 5
9) 34
10) 50
11) 27
12) 16
13) 6
14) 11
15) 12
16) 9
17) 30
18) 33
19) 75
20) 75
21) 3
22) 400
23) 7
24) 8

Division Missing Number

1) 8
2) 48
3) 9
4) 45
5) 8
6) 72
7) 8
8) 15
9) 11
10) 8
11) 8
12) 7
13) 56
14) 3
15) 18
16) 10
17) 126
18) 143
19) 10
20) 30
21) 7
22) 10
23) 1
24) 25
25) 30
26) 9

Chapter 4: Number Theory

Factoring

Factor, write prime if prime.

1) 15

2) 72

3) 25

4) 42

5) 32

6) 66

7) 34

8) 20

9) 50

10) 35

11) 40

12) 30

13) 49

14) 54

15) 96

16) 108

17) 76

18) 90

19) 100

20) 85

21) 63

22) 24

23) 48

24) 115

25) 51

26) 93

27) 52

28) 105

Prime Factorization

Factor the following numbers to their prime factors.

1.
   16
   / \
   _____

2.
   38
   / \
   _____

3.
   51
   / \
   _____

4.
   12
   / \
   _____

5.
   18
   / \
   _____

6.
   23
   / \
   _____

7.
   46
   / \
   _____

8.
   58
   / \
   _____

9.
   64
   / \
   _____

10.
    82
    / \
    _____

11.
    87
    / \
    _____

12.
    98
    / \
    _____

Divisibility Rule

Apply the divisibility rules to find the factors of each number.

1) 20   2, 3, 4, 5, 6, 9, 10   13) 24   2, 3, 4, 5, 6, 9, 10

2) 84   2, 3, 4, 5, 6, 9, 10   14) 395   2, 3, 4, 5, 6, 9, 10

3) 252   2, 3, 4, 5, 6, 9, 10   15) 920   2, 3, 4, 5, 6, 9, 10

4) 64   2, 3, 4, 5, 6, 9, 10   16) 137   2, 3, 4, 5, 6, 9, 10

5) 220   2, 3, 4, 5, 6, 9, 10   17) 440   2, 3, 4, 5, 6, 9, 10

6) 465   2, 3, 4, 5, 6, 9, 10   18) 360   2, 3, 4, 5, 6, 9, 10

7) 75   2, 3, 4, 5, 6, 9, 10   19) 495   2, 3, 4, 5, 6, 9, 10

8) 120   2, 3, 4, 5, 6, 9, 10   20) 4,870   2, 3, 4, 5, 6, 9, 10

9) 1,125   2, 3, 4, 5, 6, 9, 10   21) 590   2, 3, 4, 5, 6, 9, 10

10) 88   2, 3, 4, 5, 6, 9, 10   22) 326   2, 3, 4, 5, 6, 9, 10

11) 454   2, 3, 4, 5, 6, 9, 10   23) 114   2, 3, 4, 5, 6, 9, 10

12) 155   2, 3, 4, 5, 6, 9, 10   24) 470   2, 3, 4, 5, 6, 9, 10

Great Common Factor (GCF)

Find the GCF of the numbers.

1) 8, 26

2) 16, 44

3) 28, 38

4) 10, 35

5) 18, 48

6) 36, 52

7) 40, 75

8) 50, 45

9) 52, 8

10) 55, 85

11) 74, 94

12) 65, 20

13) 90, 10

14) 12, 34

15) 58, 86

16) 40, 95

17) 14, 56

18) 70, 100, 30

19) 64, 108

20) 63, 91

21) 20, 15, 35

22) 6, 12, 36

23) 25, 35, 70

24) 41, 39

Least Common Multiple (LCM)

Find the LCM of each.

1) 6, 14

2) 12, 18

3) 4, 16, 12

4) 10, 8

5) 10, 2, 15

6) 35, 7

7) 14, 35, 21

8) 8, 7

9) 11, 22, 44

10) 42, 21

11) 24, 72

12) 100, 25

13) 10, 5, 20

14) 15, 60

15) 20, 4, 3

16) 21, 14

17) 34, 17

18) 16, 64

19) 20, 70

20) 9, 39

21) 13, 8

22) 7, 20

23) 45, 63

24) 27, 4

Distributive Property

Multiply using the distributive property.

1) $3(x + 3) =$ _____

2) $4(x + 11) =$ _____

3) $(x + 9)5 =$ _____

4) $7(x + 6) =$ _____

5) $8(x + 8) =$ _____

6) $10(x + 4) =$ _____

7) $9(x + 10) =$ _____

8) $4(x + 8) =$ _____

9) $11(x + 6) =$ _____

10) $(x + 7)6 =$ _____

11) $(x + 13)4 =$ _____

12) $3(x + 12) =$ _____

13) $2(9x - 4) =$ _____

14) $7(8x - 3) =$ _____

15) $8(9x - 5) =$ _____

16) $(4x - 2)3 =$ _____

17) $(7x - 2)7 =$ _____

18) $(2x - 3)12 =$ _____

19) $3(4x - 1) =$ _____

20) $(-2)(4x - 3) =$ _____

21) $(-5)(x - 9) =$ _____

22) $(-7)(3x - 1) =$ _____

23) $(5x + 2)(-9) =$ _____

24) $(x + 6)(-12) =$ _____

Answer key Chapter 4

Factoring

1) 1, 3, 5, 15
2) 1, 2, 3, 4, 6, 8, 9, 12, 18, 24, 36, 72
3) 1, 5, 25
4) 1, 2, 3, 6, 7, 14, 21, 42
5) 1, 2, 4, 8, 16, 32
6) 1, 2, 3, 6, 11, 22, 33, 66
7) 1, 2, 17, 34
8) 1, 2, 4, 5, 10, 20
9) 1, 2, 5, 10, 25, 50
10) 1, 5, 7, 35
11) 1, 2, 4, 5, 8, 10, 20, 40
12) 1, 2, 3, 5, 6, 10, 15, 30
13) 1, 7, 49
14) 1, 2, 3, 6, 9, 18, 27, 54
15) 1, 2, 3, 4, 6, 8, 12, 16, 24, 32, 48, 96
16) 1, 2, 3, 4, 6, 9, 12, 18, 27, 36, 54, 108
17) 1, 2, 4, 19, 38, 76
18) 1, 2, 3, 5, 6, 9, 10, 15, 18, 30, 45, 90
19) 1, 2, 4, 5, 10, 20, 25, 50, 100
20) 1, 5, 17, 85
21) 1, 3, 7, 9, 21, 63
22) 1, 2, 3, 4, 6, 8, 12, 24
23) 1, 2, 3, 4, 6, 8, 12, 16, 48
24) 1, 5, 23, 115
25) 1, 3, 17, 51
26) 1, 3, 31, 93
27) 1, 2, 4, 13, 26, 52
28) 1, 3, 5, 7, 15, 21, 35, 105

Prime Factorization

1) $2 \times 2 \times 2 \times 2$
2) 2×19
3) 3×17
4) $2 \times 2 \times 3$
5) $2 \times 3 \times 3$
6) 23 is a prime number
7) 2×23
8) 2×29
9) $2 \times 2 \times 2 \times 2 \times 2 \times 2$
10) 2×41
11) 3×29
12) $2 \times 7 \times 7$

Divisibility Rule

1) 20 — 2, 3, <u>4</u>, <u>5</u>, 6, 9, 10
2) 84 — <u>2</u>, <u>3</u>, <u>4</u>, 5, <u>6</u>, 9, 10
3) 252 — <u>2</u>, <u>3</u>, <u>4</u>, 5, <u>6</u>, <u>9</u>, 10
4) 64 — <u>2</u>, 3, <u>4</u>, 5, 6, 9, 10
5) 220 — <u>2</u>, 3, <u>4</u>, <u>5</u>, 6, 9, <u>10</u>
6) 465 — 2, <u>3</u>, 4, <u>5</u>, 6, 9, 10
7) 75 — 2, <u>3</u>, 4, <u>5</u>, 6, 9, 10
8) 120 — <u>2</u>, <u>3</u>, <u>4</u>, <u>5</u>, <u>6</u>, 9, <u>10</u>
9) 1,125 — 2, <u>3</u>, 4, <u>5</u>, 6, <u>9</u>, 10
10) 88 — <u>2</u>, 3, <u>4</u>, 5, 6, 9, 10
11) 454 — <u>2</u>, 3, 4, 5, 6, 9, 10
12) 155 — 2, 3, 4, <u>5</u>, 6, 9, 10

PSSA Math Practice Grade 4

13) 24   <u>2</u>, <u>3</u>, <u>4</u>, 5, <u>6</u>, 9, 10
14) 395   2, 3, 4, <u>5</u>, 6, 9, 10
15) 920   <u>2</u>, 3, <u>4</u>, <u>5</u>, 6, 9, <u>10</u>
16) 137   2, 3, 4, 5, 6, 9, 10
17) 440   <u>2</u>, 3, <u>4</u>, <u>5</u>, 6, 9, <u>10</u>
18) 360   <u>2</u>, <u>3</u>, <u>4</u>, <u>5</u>, <u>6</u>, <u>9</u>, <u>10</u>
19) 495   2, <u>3</u>, 4, <u>5</u>, 6, <u>9</u>, 10
20) 4,870   <u>2</u>, 3, 4, <u>5</u>, 6, 9, <u>10</u>
21) 590   <u>2</u>, 3, 4, <u>5</u>, 6, 9, <u>10</u>
22) 326   <u>2</u>, 3, 4, 5, 6, 9, 10
23) 114   <u>2</u>, <u>3</u>, 4, 5, <u>6</u>, 9, 10
24) 470   <u>2</u>, 3, 4, <u>5</u>, 6, 9, <u>10</u>

Great Common Factor (GCF)

1) 2
2) 4
3) 2
4) 5
5) 6
6) 4
7) 5
8) 5
9) 4
10) 5
11) 2
12) 5
13) 10
14) 2
15) 2
16) 5
17) 14
18) 10
19) 4
20) 7
21) 5
22) 6
23) 5
24) 1

Least Common Multiple (LCM)

1) 42
2) 36
3) 48
4) 40
5) 30
6) 35
7) 210
8) 56
9) 44
10) 42
11) 72
12) 100
13) 20
14) 60
15) 60
16) 42
17) 34
18) 64
19) 140
20) 117
21) 104
22) 140
23) 315
24) 108

Distributive Property

1) $3x + 9$
2) $4x + 44$
3) $5x + 45$
4) $7x + 42$
5) $8x + 64$
6) $10x + 40$
7) $9x + 90$
8) $4x + 32$
9) $11x + 66$
10) $6x + 42$
11) $4x + 52$
12) $3x + 36$
13) $18x - 8$
14) $56x - 21$
15) $72x - 40$
16) $12x - 6$
17) $49x - 14$
18) $24x - 36$

19) $12x - 3$
20) $-8x + 6$
21) $-5x + 45$
22) $-21x + 7$
23) $-45x - 18$
24) $-12x - 72$

Chapter 5:

Patterns

Repeating Pattern

Circle the picture that comes next in each picture pattern.

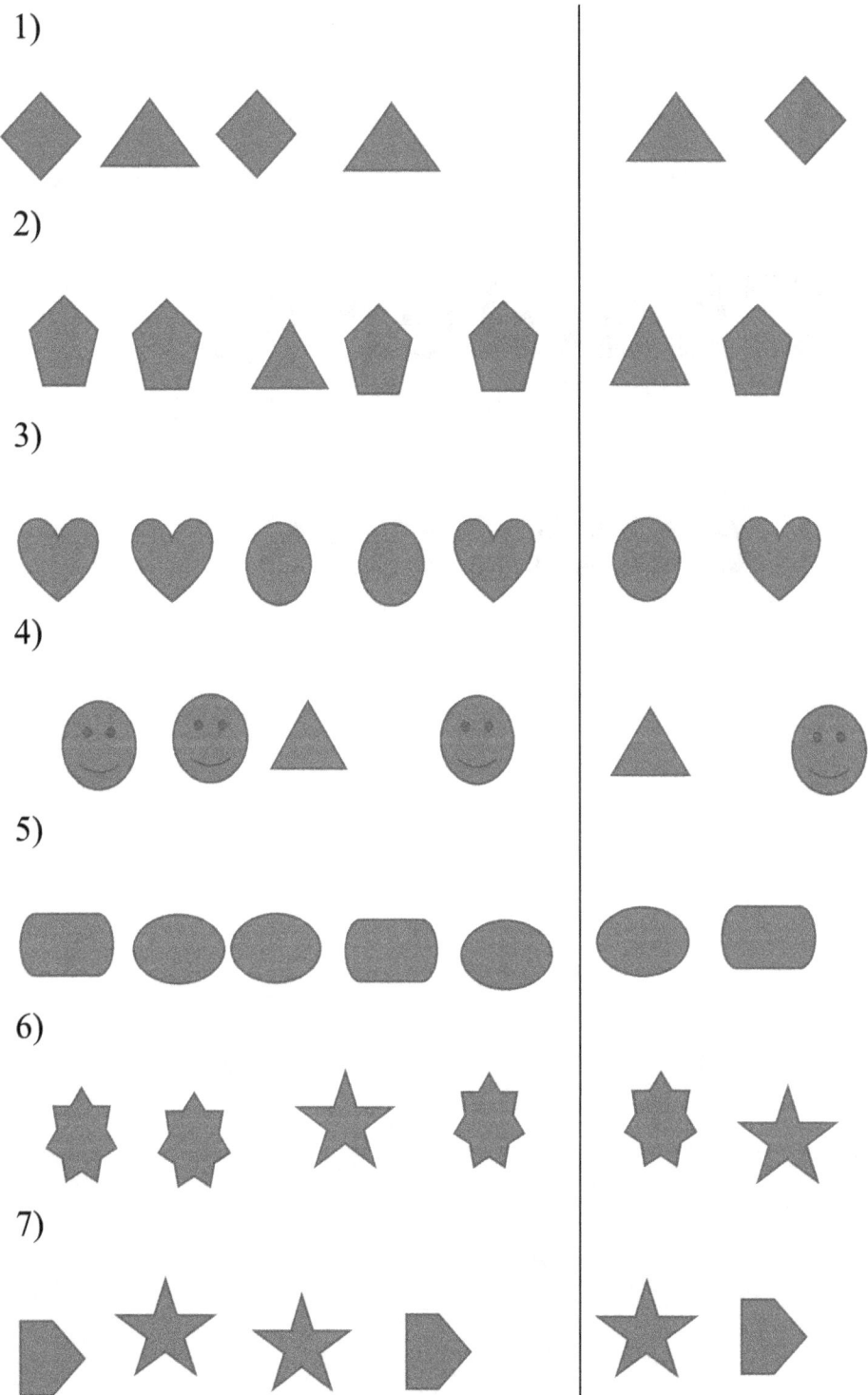

1)

2)

3)

4)

5)

6)

7)

Growing Patterns

Draw the picture that comes next in each growing pattern.

1)

2)

3)

4)

5)

6)

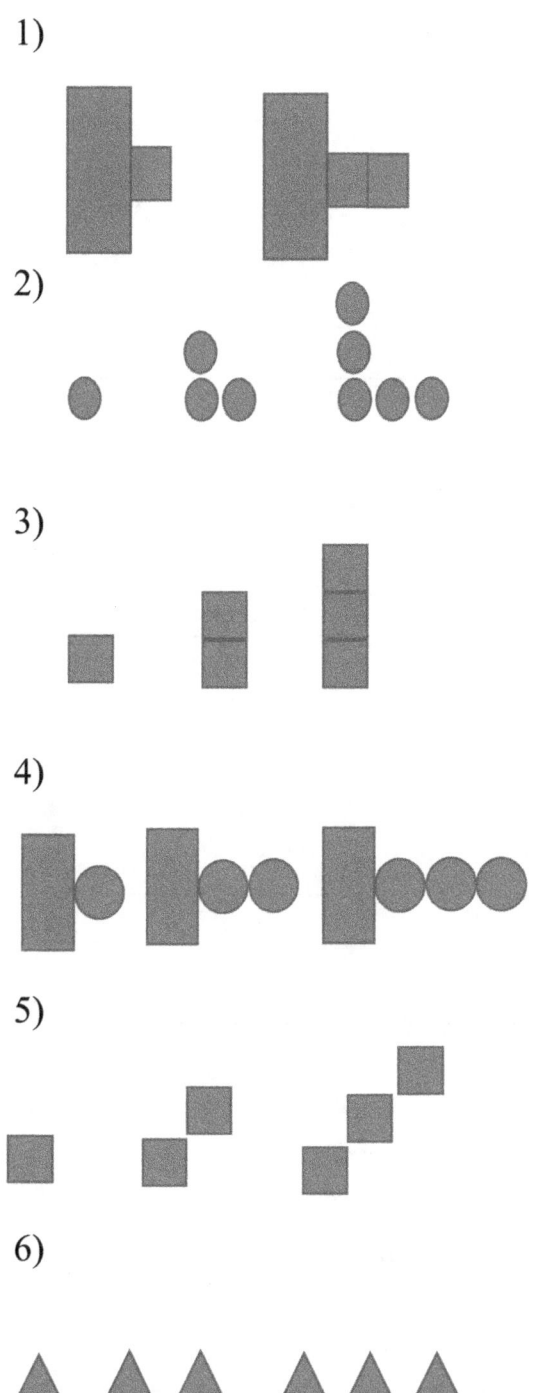

Patterns: Numbers

Continue this pattern for four more numbers:

12) 1,500; 1,350; 1,200; 1,050; _____

13) 2,800; 2,600; 2,400; 2,200; _____

14) 3,500; 3,150; 2,800; 2,450; _____

15) 1,900; 1,780; 1,660; 1,540; _____

16) 3,200; 2,950; 2,700; 2,450; _____

17) 4,100; 3,800; 3,500; 3,200; _____

18) 5,400; 4,950; 4,500; 4,050; _____

19) 2,900; 2,725; 2,550; 2,375; _____

20) 1,950; 1,700; 1,450; 1,200; _____

21) 5,500; 4,900; 4,300; 3,700; _____

22) Write a list of five numbers that follows this pattern: Start at 100 and add 400 each time.

_____

Find a Rule

Complete the output.

1- **Rule:** the output is $x + 25$

Input	x	8	15	20	38	40
Output	y					

2- **Rule:** the output is $x \times 18$

Input	x	3	7	10	11	15
Output	y					

3- **Rule:** the output is $x \div 7$

Input	x	126	147	105	280	455
Output	y					

Find a rule to write an expression.

4- **Rule:** _____

Input	x	11	13	15	20
Output	y	55	65	75	100

5- **Rule:** _____

Input	x	10	28	32	46
Output	y	14	32	36	50

6- **Rule:** _____

Input	x	84	132	180	252
Output	y	14	22	30	42

Algebraic Thinking

Circle the number sentence that fits the problem. Then solve for x.

1) Mary had $42. Then she earned more money (x). Now she has $86.

   $42 + x = $86   OR $42 + $86 = x

   x = ____

2) Lisa had $35. Then she earned more money (x). Now she has $78.

   $35 + x = $78   OR $35 + $78 = x

   x = ____

3) Matthew had $37. Then he earned more money (x). Now he has $98.

   $37 + x = $98   OR $37 + $98 = x

   x = ____

4) Charlotte gave 19 of the cookies he had baked to a friend and now he has 45 cookies left.   45 - 19 = x   OR   x - 19 = 45

   x = ____

5) Mia gave 32 of the cookies she had baked to a friend and now she has 55 cookies left.   55 - 32 = x   OR   x - 32 = 55

   x = ____

6) Lucas gave 41 of the cookies he had baked to a friend and now he has 49 cookies left. .   49 - 41 = x   OR   x - 41 = 49

   x = ____

Answers of Worksheets – Chapter 5

Repeating pattern

1)

2)

3)

4)

5)

6)

7)

Growing patterns

1)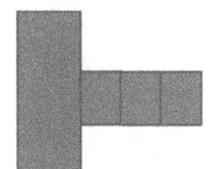

2)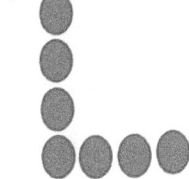

3)

4)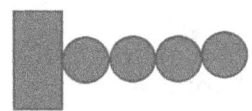

5)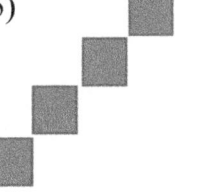

6)

Pattern

12) 900; 750; 600; 450

13) 2,000; 1,800; 1,600; 1,400

14) 2,100; 1,750; 1,400; 1,050

15) 1,420; 1,300; 1,180; 1,060

16) 2,200; 1,950; 1,700; 1,450

17) 2,900; 2,600; 2,300; 2,000

18) 3,600; 3,150; 2,700; 2,250

19) 2,200; 2,025; 1,850; 1,675

20) 950; 700; 450; 200

22) 100; 500; 900; 1,300; 1,700

21) 3,100; 2,500; 1,900; 1,300

Find a Rule

1)
Input	x	8	15	20	38	40
Output	y	33	40	45	63	65

2)
Input	x	3	7	10	11	15
Output	y	54	126	180	198	270

3)
Input	x	126	147	105	280	455
Output	y	18	21	15	40	65

4) $y = 5x$ 5) $y = x + 4$ 6) $y = x \div 6$

Algebraic Thinking

1) $\$42 + x = \$86$; $x = 44$

2) $\$35 + x = \$78$; $x = 43$

3) $\$37 + x = \$98$; $x = 61$

4) $x - 19 = 45$; $x = 64$

5) $x - 32 = 55$; $x = 87$

6) $x - 41 = 49$; $x = 90$

Chapter 6: Fractions and Mix Numbers

Adding Fractions – Like Denominator

Find each sum.

1) $\dfrac{1}{3} + \dfrac{1}{3} =$

2) $\dfrac{3}{7} + \dfrac{1}{7} =$

3) $\dfrac{2}{9} + \dfrac{5}{9} =$

4) $\dfrac{7}{15} + \dfrac{1}{15} =$

5) $\dfrac{5}{23} + \dfrac{4}{23} =$

6) $\dfrac{8}{29} + \dfrac{7}{29} =$

7) $\dfrac{7}{19} + \dfrac{1}{19} =$

8) $\dfrac{5}{16} + \dfrac{1}{16} =$

9) $\dfrac{5}{31} + \dfrac{8}{31} =$

10) $\dfrac{5}{51} + \dfrac{8}{51} =$

11) $\dfrac{1}{17} + \dfrac{3}{17} =$

12) $\dfrac{2}{13} + \dfrac{4}{13} =$

13) $\dfrac{5}{41} + \dfrac{19}{41} =$

14) $\dfrac{2}{55} + \dfrac{9}{55} =$

15) $\dfrac{5}{21} + \dfrac{8}{21} =$

16) $\dfrac{10}{33} + \dfrac{2}{33} =$

17) $\dfrac{3}{11} + \dfrac{3}{11} =$

18) $\dfrac{24}{67} + \dfrac{1}{67} =$

19) $\dfrac{3}{19} + \dfrac{6}{19} =$

20) $\dfrac{20}{47} + \dfrac{13}{47} =$

Adding Fractions – Unlike Denominator

Add the fractions and simplify the answers.

1) $\frac{1}{2}+\frac{1}{6}=$

2) $\frac{2}{3}+\frac{3}{4}=$

3) $\frac{3}{5}+\frac{1}{2}=$

4) $\frac{7}{10}+\frac{1}{3}=$

5) $\frac{4}{15}+\frac{1}{5}=$

6) $\frac{3}{14}+\frac{2}{7}=$

7) $\frac{2}{7}+\frac{1}{3}=$

8) $\frac{1}{20}+\frac{1}{5}=$

9) $\frac{7}{12}+\frac{1}{6}=$

10) $\frac{1}{8}+\frac{3}{4}=$

11) $\frac{4}{21}+\frac{1}{3}=$

12) $\frac{5}{36}+\frac{2}{9}=$

13) $\frac{4}{35}+\frac{3}{7}=$

14) $\frac{2}{33}+\frac{1}{11}=$

15) $\frac{10}{27}+\frac{1}{3}=$

16) $\frac{7}{45}+\frac{4}{9}=$

17)

18) $\frac{1}{7}+\frac{2}{5}=$

19) $\frac{1}{2}+\frac{5}{22}=$

20) $\frac{3}{16}+\frac{1}{4}=$

21) $\frac{5}{16}+\frac{1}{24}=$

22) $\frac{2}{15}+\frac{3}{10}=$

23) $\frac{5}{42}+\frac{4}{21}=$

24) $\frac{1}{36}+\frac{5}{24}=$

Subtracting Fractions – Like Denominator

Find the difference.

1) $\dfrac{6}{5} - \dfrac{1}{5} =$

2) $\dfrac{7}{12} - \dfrac{4}{12} =$

3) $\dfrac{8}{13} - \dfrac{5}{13} =$

4) $\dfrac{19}{6} - \dfrac{7}{6} =$

5) $\dfrac{11}{21} - \dfrac{9}{21} =$

6) $\dfrac{15}{43} - \dfrac{7}{43} =$

7) $\dfrac{9}{23} - \dfrac{3}{23} =$

8) $\dfrac{18}{47} - \dfrac{15}{47} =$

9) $\dfrac{8}{20} - \dfrac{4}{20} =$

10) $\dfrac{34}{48} - \dfrac{17}{48} =$

11) $\dfrac{6}{7} - \dfrac{2}{7} =$

12) $\dfrac{36}{51} - \dfrac{28}{51} =$

13) $\dfrac{9}{11} - \dfrac{5}{11} =$

14) $\dfrac{29}{49} - \dfrac{14}{49} =$

15) $\dfrac{12}{17} - \dfrac{6}{17} =$

16) $\dfrac{17}{23} - \dfrac{11}{23} =$

17) $\dfrac{5}{7} - \dfrac{1}{7} =$

18) $\dfrac{12}{31} - \dfrac{8}{31} =$

19) $\dfrac{38}{61} - \dfrac{29}{61} =$

20) $\dfrac{31}{52} - \dfrac{20}{52} =$

21) $\dfrac{51}{63} - \dfrac{46}{63} =$

22) $\dfrac{55}{83} - \dfrac{25}{83} =$

23) $\dfrac{54}{77} - \dfrac{30}{77} =$

24) $\dfrac{49}{55} - \dfrac{37}{55} =$

Subtracting Fractions – Unlike Denominator

Solve each problem.

1) $\frac{1}{3} - \frac{1}{6} =$

2) $\frac{3}{5} - \frac{1}{9} =$

3) $\frac{1}{4} - \frac{1}{6} =$

4) $\frac{7}{9} - \frac{3}{10} =$

5) $\frac{11}{12} - \frac{5}{24} =$

6) $\frac{9}{16} - \frac{3}{20} =$

7) $\frac{19}{30} - \frac{1}{6} =$

8) $\frac{1}{2} - \frac{7}{15} =$

9) $\frac{3}{5} - \frac{2}{7} =$

10) $\frac{8}{9} - \frac{4}{11} =$

11) $\frac{6}{8} - \frac{7}{48} =$

12) $\frac{3}{4} - \frac{7}{10} =$

13) $\frac{4}{5} - \frac{8}{45} =$

14) $\frac{6}{7} - \frac{3}{10} =$

15) $\frac{11}{12} - \frac{13}{24} =$

16) $\frac{4}{9} - \frac{15}{63} =$

17) $\frac{5}{12} - \frac{5}{16} =$

18) $\frac{5}{8} - \frac{3}{10} =$

19) $\frac{5}{7} - \frac{3}{8} =$

20) $\frac{3}{4} - \frac{21}{44} =$

Converting Mix Numbers

Convert the following mixed numbers into improper fractions.

1) $2\frac{3}{5} =$

2) $3\frac{4}{5} =$

3) $5\frac{3}{4} =$

4) $3\frac{5}{8} =$

5) $6\frac{2}{5} =$

6) $7\frac{8}{9} =$

7) $2\frac{2}{15} =$

8) $3\frac{1}{13} =$

9) $2\frac{1}{12} =$

10) $5\frac{5}{6} =$

11) $7\frac{3}{5} =$

12) $3\frac{9}{10} =$

13) $6\frac{1}{3} =$

14) $5\frac{5}{6} =$

15) $8\frac{2}{5} =$

16) $4\frac{3}{8} =$

17) $3\frac{5}{9} =$

18) $2\frac{3}{14} =$

19) $7\frac{5}{6} =$

20) $5\frac{6}{7} =$

21) $4\frac{7}{8} =$

22) $3\frac{2}{9} =$

23) $2\frac{9}{11} =$

24) $10\frac{5}{7} =$

Converting improper Fractions

Convert the following improper fractions into mixed numbers

1) $\frac{55}{16}=$

2) $\frac{95}{34}=$

3) $\frac{39}{19}=$

4) $\frac{11}{3}=$

5) $\frac{61}{17}=$

6) $\frac{152}{41}=$

7) $\frac{125}{31}=$

8) $\frac{46}{7}=$

9) $\frac{43}{11}=$

10) $\frac{14}{3}=$

11) $\frac{29}{6}=$

12) $\frac{61}{15}=$

13) $\frac{56}{24}=$

14) $\frac{21}{9}=$

15) $\frac{115}{16}=$

16) $\frac{59}{6}=$

17) $\frac{144}{10}=$

18) $\frac{51}{13}=$

19) $\frac{36}{8}=$

20) $\frac{58}{6}=$

21) $\frac{9}{7}=$

22) $\frac{89}{11}=$

23) $\frac{102}{9}=$

24) $\frac{180}{19}=$

Adding Mix Numbers

Add the following fractions.

1) $3\frac{4}{9} + 2\frac{2}{9} =$

2) $3\frac{2}{5} + 2\frac{1}{5} =$

3) $1\frac{1}{7} + 2\frac{3}{7} =$

4) $4\frac{5}{6} + 2\frac{1}{3} =$

5) $1\frac{4}{15} + 2\frac{2}{5} =$

6) $4\frac{1}{3} + 1\frac{3}{4} =$

7) $3\frac{7}{9} + 3\frac{1}{6} =$

8) $3\frac{5}{8} + 2\frac{1}{2} =$

9) $3\frac{3}{4} + 2\frac{1}{4} =$

10) $1\frac{4}{11} + 2\frac{3}{11} =$

11) $4\frac{1}{2} + 2\frac{2}{5} =$

12) $5\frac{1}{4} + 2\frac{5}{6} =$

13) $6\frac{2}{7} + 2\frac{5}{7} =$

14) $3\frac{7}{8} + 2\frac{3}{16} =$

15) $3\frac{3}{4} + 3\frac{2}{9} =$

16) $4\frac{3}{5} + 2\frac{1}{6} =$

17) $7\frac{1}{2} + 6\frac{3}{5} =$

18) $6\frac{2}{3} + 1\frac{5}{12} =$

19) $2\frac{1}{9} + 7\frac{2}{3} =$

20) $4\frac{1}{6} + 2\frac{5}{9} =$

21) $5\frac{2}{3} + 6\frac{3}{4} =$

22) $7\frac{1}{8} + 1\frac{7}{24} =$

23) $5\frac{3}{7} + 4\frac{1}{8} =$

24) $8\frac{1}{3} + 4\frac{3}{4} =$

Subtracting Mix Numbers

Subtract the following fractions.

1) $8\frac{1}{4} - 7\frac{1}{4} =$

2) $5\frac{5}{6} - 5\frac{1}{6} =$

3) $9\frac{7}{11} - 8\frac{3}{11} =$

4) $5\frac{1}{2} - 2\frac{1}{6} =$

5) $4\frac{1}{4} - 1\frac{1}{8} =$

6) $9\frac{1}{3} - 5\frac{3}{7} =$

7) $5\frac{7}{9} - 2\frac{2}{9} =$

8) $8\frac{15}{17} - 5\frac{11}{17} =$

9) $9\frac{11}{14} - 3\frac{5}{14} =$

10) $7\frac{9}{10} - 6\frac{7}{10} =$

11) $8\frac{3}{4} - 5\frac{1}{12} =$

12) $6\frac{7}{8} - 3\frac{1}{8} =$

13) $7\frac{12}{35} - 3\frac{3}{7} =$

14) $6\frac{1}{3} - 4\frac{1}{9} =$

15) $10\frac{6}{7} - 7\frac{3}{7} =$

16) $9\frac{2}{3} - 3\frac{1}{3} =$

17) $5\frac{4}{11} - 3\frac{2}{11} =$

18) $7\frac{3}{10} - 5\frac{1}{5} =$

19) $8\frac{5}{6} - 5\frac{1}{12} =$

20) $3\frac{3}{4} - 3\frac{7}{16} =$

21) $8\frac{8}{13} - 3\frac{1}{3} =$

22) $6\frac{5}{6} - 4\frac{7}{30} =$

23) $5\frac{6}{7} - 4\frac{4}{11} =$

24) $6\frac{10}{19} - 2\frac{9}{19} =$

Simplify Fractions

Reduce these fractions to lowest terms

1) $\frac{18}{12} =$

2) $\frac{22}{33} =$

3) $\frac{32}{40} =$

4) $\frac{27}{36} =$

5) $\frac{8}{24} =$

6) $\frac{15}{35} =$

7) $\frac{20}{35} =$

8) $\frac{56}{70} =$

9) $\frac{9}{81} =$

10) $\frac{40}{16} =$

11) $\frac{54}{72} =$

12) $\frac{40}{120} =$

13) $\frac{12}{20} =$

14) $\frac{7}{28} =$

15) $\frac{14}{49} =$

16) $\frac{58}{87} =$

17) $\frac{72}{27} =$

18) $\frac{48}{180} =$

19) $\frac{24}{64} =$

20) $\frac{48}{42} =$

21) $\frac{120}{240} =$

22) $\frac{54}{279} =$

23) $\frac{340}{68} =$

24) $\frac{150}{600} =$

Multiplying Fractions

Find the product.

1) $\frac{2}{3} \times \frac{5}{7} =$

2) $\frac{3}{11} \times \frac{4}{9} =$

3) $\frac{7}{24} \times \frac{3}{14} =$

4) $\frac{7}{16} \times \frac{24}{35} =$

5) $\frac{15}{21} \times \frac{3}{5} =$

6) $\frac{18}{20} \times \frac{4}{9} =$

7) $\frac{6}{7} \times \frac{7}{9} =$

8) $\frac{54}{79} \times 0 =$

9) $\frac{2}{6} \times \frac{12}{14} =$

10) $\frac{24}{14} \times \frac{7}{8} =$

11) $\frac{38}{36} \times \frac{18}{19} =$

12) $\frac{8}{10} \times \frac{5}{64} =$

13) $\frac{15}{4} \times \frac{16}{9} =$

14) $\frac{25}{8} \times \frac{4}{10} =$

15) $\frac{14}{63} \times \frac{9}{7} =$

16) $\frac{12}{20} \times 4 =$

17) $\frac{7}{33} \times \frac{66}{21} =$

18) $\frac{5}{16} \times \frac{8}{10} =$

19) $\frac{9}{10} \times \frac{4}{27} =$

20) $\frac{6}{42} \times \frac{7}{12} =$

21) $\frac{8}{19} \times \frac{1}{16} =$

22) $\frac{10}{7} \times \frac{4}{80} =$

23) $\frac{9}{12} \times \frac{4}{54} =$

24) $\frac{60}{400} \times \frac{200}{600} =$

Multiplying Mixed Number

Multiply. Reduce to lowest terms.

1) $3\frac{1}{4} \times 3\frac{1}{5} =$

2) $2\frac{2}{7} \times 1\frac{1}{8} =$

3) $1\frac{1}{4} \times 2\frac{3}{5} =$

4) $2\frac{2}{9} \times 1\frac{1}{10} =$

5) $3\frac{3}{5} \times 2\frac{1}{5} =$

6) $2\frac{3}{4} \times 2\frac{2}{3} =$

7) $4\frac{1}{2} \times 1\frac{1}{9} =$

8) $2\frac{4}{5} \times 4\frac{1}{7} =$

9) $3\frac{1}{4} \times 2\frac{1}{3} =$

10) $1\frac{1}{5} \times 5\frac{1}{6} =$

11) $5\frac{1}{3} \times 2\frac{1}{8} =$

12) $2\frac{1}{3} \times 1\frac{2}{9} =$

13) $3\frac{1}{4} \times 2\frac{2}{5} =$

14) $4\frac{1}{9} \times 3\frac{1}{3} =$

15) $3\frac{1}{5} \times 2\frac{1}{7} =$

16) $4\frac{1}{2} \times 2\frac{2}{5} =$

17) $1\frac{1}{4} \times 2\frac{4}{5} =$

18) $3\frac{1}{3} \times 1\frac{1}{4} =$

19) $4\frac{4}{5} \times 1\frac{5}{7} =$

20) $6\frac{3}{4} \times 1\frac{1}{3} =$

21) $5\frac{1}{2} \times 3\frac{1}{5} =$

22) $4\frac{2}{3} \times 6\frac{1}{2} =$

Comparing Fractions

Compare the fractions, and write >, < or =

1) $\frac{13}{2}$ ___ $\frac{19}{10}$

2) $\frac{15}{4}$ ___ $\frac{5}{7}$

3) $\frac{5}{8}$ ___ $\frac{4}{5}$

4) $\frac{11}{3}$ ___ $\frac{12}{8}$

5) $\frac{2}{9}$ ___ $\frac{4}{7}$

6) $\frac{13}{5}$ ___ $\frac{17}{4}$

7) $\frac{14}{9}$ ___ $\frac{8}{11}$

8) $\frac{15}{13}$ ___ $\frac{21}{8}$

9) $5\frac{1}{10}$ ___ $8\frac{1}{15}$

10) $9\frac{1}{12}$ ___ $7\frac{1}{9}$

11) $4\frac{1}{4}$ ___ $4\frac{1}{7}$

12) $8\frac{6}{7}$ ___ $8\frac{3}{8}$

13) $1\frac{5}{9}$ ___ $4\frac{2}{3}$

14) $\frac{1}{24}$ ___ $\frac{2}{13}$

15) $\frac{42}{23}$ ___ $\frac{29}{72}$

16) $\frac{14}{200}$ ___ $\frac{8}{81}$

17) $19\frac{1}{3}$ ___ $19\frac{1}{7}$

18) $\frac{1}{8}$ ___ $\frac{1}{12}$

19) $\frac{1}{11}$ ___ $\frac{1}{15}$

20) $\frac{1}{15}$ ___ $\frac{7}{12}$

21) $\frac{10}{33}$ ___ $\frac{8}{59}$

22) $\frac{6}{7}$ ___ $\frac{3}{8}$

23) $6\frac{2}{5}$ ___ $4\frac{12}{5}$

24) $2\frac{12}{5}$ ___ $3\frac{4}{5}$

Answer key Chapter 6

Adding Fractions – Like Denominator

1) $\frac{2}{3}$
2) $\frac{4}{7}$
3) $\frac{7}{9}$
4) $\frac{8}{15}$
5) $\frac{9}{23}$
6) $\frac{15}{29}$
7) $\frac{8}{19}$
8) $\frac{3}{8}$
9) $\frac{13}{31}$
10) $\frac{13}{51}$
11) $\frac{4}{17}$
12) $\frac{6}{13}$
13) $\frac{24}{41}$
14) $\frac{1}{5}$
15) $\frac{13}{21}$
16) $\frac{4}{11}$
17) $\frac{6}{11}$
18) $\frac{25}{67}$
19) $\frac{9}{19}$
20) $\frac{33}{47}$

Adding Fractions – Unlike Denominator

1) $\frac{2}{3}$
2) $\frac{17}{12}$
3) $\frac{11}{10}$
4) $\frac{31}{30}$
5) $\frac{7}{15}$
6) $\frac{1}{2}$
7) $\frac{13}{21}$
8) $\frac{1}{4}$
9) $\frac{3}{4}$
10) $\frac{7}{8}$
11) $\frac{11}{21}$
12) $\frac{13}{36}$
13) $\frac{19}{35}$
14) $\frac{5}{33}$
15) $\frac{19}{27}$
16) $\frac{3}{5}$
17) $\frac{19}{45}$
18) $\frac{19}{35}$
19) $\frac{8}{11}$
20) $\frac{7}{16}$
21) $\frac{17}{48}$
22) $\frac{13}{30}$
23) $\frac{13}{42}$
24) $\frac{17}{72}$

Subtracting Fractions – Like Denominator

1) 1
2) $\frac{1}{4}$
3) $\frac{3}{13}$
4) 2
5) $\frac{2}{21}$
6) $\frac{8}{43}$
7) $\frac{6}{23}$
8) $\frac{3}{47}$
9) $\frac{1}{5}$
10) $\frac{17}{48}$
11) $\frac{4}{7}$
12) $\frac{8}{51}$
13) $\frac{4}{11}$
14) $\frac{15}{49}$
15) $\frac{6}{17}$
16) $\frac{6}{23}$
17) $\frac{4}{7}$
18) $\frac{4}{31}$

PSSA Math Practice Grade 4

19) $\frac{9}{61}$

20) $\frac{11}{52}$

21) $\frac{5}{63}$

22) $\frac{30}{83}$

23) $\frac{24}{77}$

24) $\frac{12}{55}$

Subtracting Fractions – Unlike Denominator

1) $\frac{1}{6}$

2) $\frac{22}{45}$

3) $\frac{1}{12}$

4) $\frac{43}{90}$

5) $\frac{17}{24}$

6) $\frac{33}{80}$

7) $\frac{7}{15}$

8) $\frac{1}{30}$

9) $\frac{11}{35}$

10) $\frac{52}{99}$

11) $\frac{29}{48}$

12) $\frac{1}{20}$

13) $\frac{28}{45}$

14) $\frac{39}{70}$

15) $\frac{3}{8}$

16) $\frac{13}{63}$

17) $\frac{5}{48}$

18) $\frac{13}{40}$

19) $\frac{19}{56}$

20) $\frac{3}{11}$

Converting Mix Numbers

1) $\frac{13}{5}$

2) $\frac{19}{5}$

3) $\frac{23}{4}$

4) $\frac{29}{8}$

5) $\frac{32}{5}$

6) $\frac{71}{9}$

7) $\frac{32}{15}$

8) $\frac{40}{13}$

9) $\frac{25}{12}$

10) $\frac{35}{6}$

11) $\frac{38}{5}$

12) $\frac{39}{10}$

13) $\frac{19}{3}$

14) $\frac{35}{6}$

15) $\frac{42}{5}$

16) $\frac{35}{8}$

17) $\frac{32}{9}$

18) $\frac{31}{14}$

19) $\frac{47}{6}$

20) $\frac{41}{7}$

21) $\frac{39}{8}$

22) $\frac{29}{9}$

23) $\frac{31}{11}$

24) $\frac{75}{7}$

Converting improper Fractions

1) $3\frac{7}{16}$

2) $2\frac{27}{34}$

3) $2\frac{1}{19}$

4) $3\frac{2}{3}$

5) $3\frac{10}{17}$

6) $3\frac{29}{41}$

7) $4\frac{1}{31}$

8) $6\frac{4}{7}$

9) $3\frac{10}{11}$

10) $4\frac{2}{3}$

11) $4\frac{5}{6}$

12) $4\frac{1}{15}$

13) $2\frac{1}{3}$

14) $2\frac{1}{3}$

15) $7\frac{3}{16}$

16) $9\frac{5}{6}$

17) $14\frac{2}{5}$

18) $3\frac{12}{13}$

19) $4\frac{1}{2}$

20) $9\frac{2}{3}$

21) $1\frac{2}{7}$

22) $8\frac{1}{11}$

23) $11\frac{1}{3}$

24) $9\frac{9}{19}$

Adding Mix Numbers

1) $5\frac{2}{3}$

2) $5\frac{3}{5}$

3) $3\frac{4}{7}$

4) $7\frac{1}{6}$

5) $3\frac{2}{3}$

6) $6\frac{1}{12}$

7) $6\frac{17}{18}$

8) $6\frac{1}{8}$

9) 6

10) $3\frac{7}{11}$

11) $6\frac{9}{10}$

12) $8\frac{1}{12}$

13) 9

14) $6\frac{1}{16}$

15) $6\frac{35}{36}$

16) $6\frac{23}{30}$

17) $14\frac{1}{10}$

18) $8\frac{1}{12}$

19) $9\frac{7}{9}$

20) $6\frac{13}{18}$

21) $12\frac{5}{12}$

22) $8\frac{5}{12}$

23) $9\frac{31}{56}$

24) $13\frac{1}{12}$

Subtracting Mix Numbers

1) 1

2) $\frac{2}{3}$

3) $1\frac{4}{11}$

4) $3\frac{1}{3}$

5) $3\frac{1}{8}$

6) $3\frac{19}{21}$

7) $3\frac{5}{9}$

8) $3\frac{4}{17}$

9) $6\frac{3}{7}$

10) $1\frac{1}{5}$

11) $3\frac{2}{3}$

12) $3\frac{3}{4}$

13) $3\frac{32}{35}$

14) $2\frac{2}{9}$

15) $3\frac{3}{7}$

16) $6\frac{1}{3}$

17) $2\frac{2}{11}$

18) $2\frac{1}{10}$

19) $3\frac{3}{4}$

20) $\frac{5}{16}$

21) $5\frac{11}{39}$

22) $2\frac{3}{5}$

23) $1\frac{38}{77}$

24) $4\frac{1}{19}$

Simplify Fractions

1) $\frac{3}{2}$

2) $\frac{2}{3}$

3) $\frac{4}{5}$

4) $\frac{3}{4}$

5) $\frac{1}{3}$

6) $\frac{3}{7}$

7) $\frac{4}{7}$

8) $\frac{4}{5}$

9) $\frac{1}{9}$

10) $\frac{5}{2}$

11) $\frac{3}{4}$

12) $\frac{1}{3}$

13) $\frac{3}{5}$

14) $\frac{1}{4}$

15) $\frac{2}{7}$

16) $\frac{2}{3}$

17) $\frac{8}{3}$

18) $\frac{4}{15}$

19) $\frac{3}{8}$

20) $\frac{8}{7}$

21) $\frac{1}{2}$

22) $\frac{6}{31}$

23) 5

24) $\frac{1}{4}$

Multiplying Fractions

1) $\frac{10}{21}$

2) $\frac{4}{33}$

3) $\frac{1}{16}$

4) $\frac{3}{10}$

5) $\frac{3}{7}$

6) $\frac{2}{5}$

7) $\frac{2}{3}$

8) 0

9) $\frac{2}{7}$

10) $\frac{3}{2}$

11) 1

12) $\frac{1}{16}$

13) $\frac{20}{3}$

14) $\frac{5}{4}$

15) $\frac{2}{7}$

16) $\frac{12}{5}$

17) $\frac{2}{3}$

18) $\frac{1}{4}$

19) $\frac{2}{15}$

20) $\frac{1}{12}$

21) $\frac{1}{38}$

22) $\frac{1}{14}$

23) $\frac{1}{18}$

24) $\frac{1}{20}$

Multiplying Mixed Number

1) $10\frac{2}{5}$

2) $2\frac{4}{7}$

3) $3\frac{1}{4}$

4) $2\frac{4}{9}$

5) $7\frac{23}{25}$

6) $7\frac{1}{3}$

7) 5

8) $11\frac{3}{5}$

9) $7\frac{7}{12}$

10) $6\frac{1}{5}$

11) $11\frac{1}{3}$

12) $2\frac{23}{27}$

13) $7\frac{4}{5}$

14) $13\frac{19}{27}$

15) $6\frac{6}{7}$

16) $10\frac{4}{5}$

17) $3\frac{1}{2}$

18) $4\frac{1}{6}$

19) $8\frac{8}{35}$

20) 9

21) $17\frac{3}{5}$

22) $30\frac{1}{3}$

Comparing Fractions

1) >
2) >
3) <
4) >
5) <
6) <

7) >
8) <
9) <
10) >
11) >
12) >

13) <
14) <
15) >
16) <
17) >
18) >

19) >
20) <
21) >
22) >
23) =
24) >

Chapter 7:

Decimal

Graph Decimals

Write the decimals indicated by the arrows.

1)
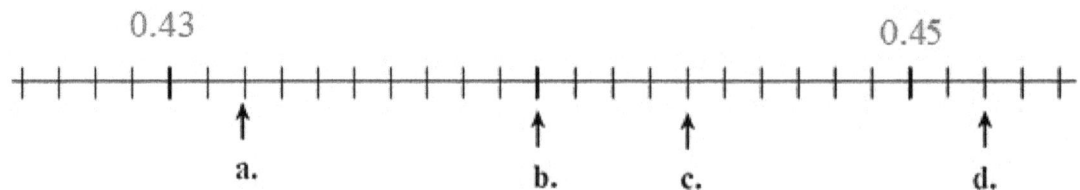

a. _____ b. _____ c. _____ d. _____

2)
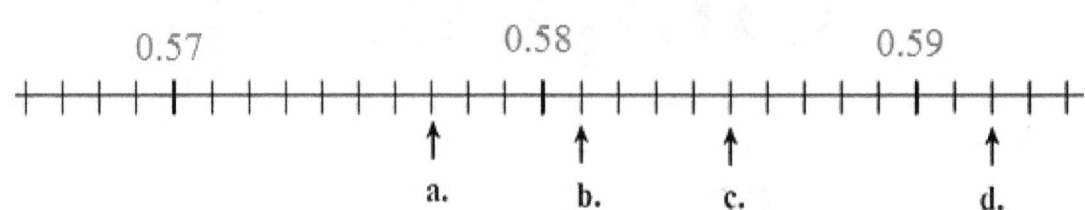

a. _____ b. _____ c. _____ d. _____

3)
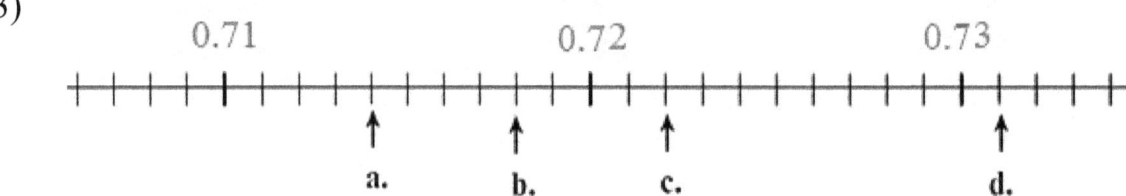

a. _____ b. _____ c. _____ d. _____

4)
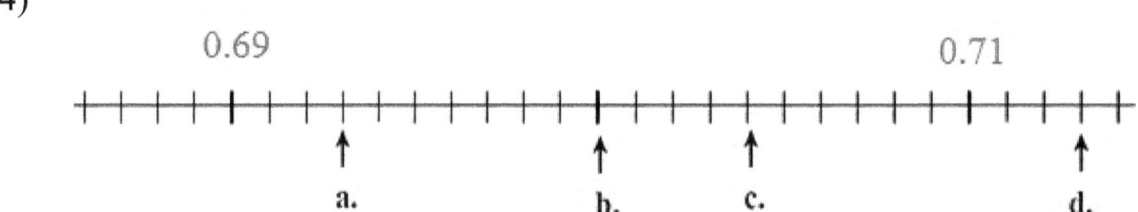

a. _____ b. _____ c. _____ d. _____

Round Decimals

Round each number to the correct place value

1) 0.9<u>2</u> =

2) 4.<u>2</u>1 =

3) 8.<u>8</u>33 =

4) 0.<u>5</u>69 =

5) <u>7</u>.779 =

6) 0.0<u>4</u>9 =

7) 9.<u>3</u>7 =

8) 25.3<u>3</u>1 =

9) 4.6<u>2</u>9 =

10) 10.<u>4</u>71 =

11) 3.<u>2</u>6 =

12) 4.<u>2</u>29 =

13) 6.<u>2</u>18 =

14) 9.1<u>6</u>24 =

15) 3<u>6</u>.29 =

16) 4<u>7</u>.68 =

17) 6<u>3</u>2.785 =

18) 577.<u>8</u>29 =

19) 24.5<u>7</u>9 =

20) 8<u>3</u>.91 =

21) 5.4<u>1</u>35 =

22) 86.<u>2</u>76 =

23) 354.<u>3</u>39 =

24) 0.9<u>2</u>66 =

25) 0.00<u>7</u>4 =

26) 8.0<u>4</u>49 =

27) 41.6<u>5</u>96 =

28) 19.0<u>8</u>27 =

Decimals Addition

Add the following.

1) 32.19 + 31.16

2) 0.54 + 0.31

3) 21.42 + 12.57

4) 57.189 + 7.221

5) 22.740 + 8.37

6) 6.712 + 4.105

7) 78.46 + 18.54

8) 66.24 + 20.36

9) 39.11 + 15.09

10) 4.88 + 19.45

11) 16.254 + 34.227

12) 36.66 + 4.82

13) 42.45 + 9.37

14) 124.41 + 5.71

Decimals Subtraction

Subtract the following

1) 8.19 − 2.46

2) 46.25 − 16.18

3) 0.89 − 0.5

4) 24.354 − 7.8

5) 0.789 − 0.06

6) 63.25 − 42.52

7) 157.75 − 94.87

8) 45.65 − 19.57

9) 65.42 − 47.71

10) 8.652 − 0.553

11) 48.61 − 29.52

12) 16.399 − 5.389

13) 35.251 − 9.169

14) 148.42 − 11.78

Decimals Multiplication

Solve.

1) 2.4 × 1.9

2) 3.6 × 5.2

3) 4.02 × 2.04

4) 45.9 × 10

5) 49.8 × 100

6) 41.56 × 4.2

7) 24.51 × 10.2

8) 1.89 × 7.35

9) 14.05 × 0.09

10) 51.03 × 4.04

11) 14.56 × 12.4

12) 9.56 × 0.04

13) 7.5 × 0.11

14) 23.1 × 4.02

Decimal Division

Dividing Decimals.

1) 7 ÷ 1,000 =

2) 3 ÷ 10,000 =

3) 2.8 ÷ 10 =

4) 0.08 ÷ 100 =

5) 5 ÷ 25 =

6) 4 ÷ 48 =

7) 8 ÷ 40 =

8) 7 ÷ 140 =

9) 9 ÷ 10,000 =

10) 0.6 ÷ 0.54 =

11) 0.4 ÷ 0.004 =

12) 0.3 ÷ 0.15 =

13) 0.7 ÷ 0.56 =

14) 0.6 ÷ 0.0006 =

15) 4.9 ÷ 100 =

16) 7.3 ÷ 100 =

17) 8.5 ÷ 10 =

18) 16.2 ÷ 4.4 =

19) 36.9 ÷ 3.3 =

20) 0.8 ÷ 0.08 =

21) 8.04 ÷ 4.2 =

22) 0.09 ÷ 0.30 =

23) 0.8 ÷ 6.4 =

24) 0.07 ÷ 42 =

25) 6.28 ÷ 0.6 =

26) 0.026 ÷ 13 =

Comparing Decimals

Write the Correct Comparison Symbol (>, < or =)

1) 1.98 ____ 3.12

2) 0.6 ____ 0.549

3) 19.01 ____ 19.010

4) 5.05 ____ 5.50

5) 0.811 ____ 0.81

6) 0.658 ____ 0.865

7) 5.46 ____ 5.391

8) 7.021 ____ 7.035

9) 56.321 ____ 56.123

10) 4.69 ____ 4.069

11) 3.55 ____ 3.555

12) 0.08 ____ 0.12

13) 2.405 ____ 2.45

14) 7.53 ____ 7.35

15) 0.11 ____ 0.011

16) 86.09 ____ 86.090

17) 0.190 ____ 0.21

18) 53.98 ____ 54.07

19) 0.072 ____ 0.720

20) 43.2 ____ 34.9

21) 17.99 ____ 20.19

22) 0.087 ____ 0.0807

23) 3.059 ____ 0.3059

24) 8.2 ____ 0.825

25) 8.77 ____ 0.877

26) 9.56 ____ 9.5600

27) 4.97 ____ 0.497

28) 3.0504 ____ 3.0540

Convert Fraction to Decimal

Write each as a decimal.

1) $\frac{6}{10} =$

2) $\frac{58}{100} =$

3) $\frac{80}{100} =$

4) $\frac{5}{40} =$

5) $\frac{4}{50} =$

6) $\frac{7}{100} =$

7) $\frac{5}{80} =$

8) $\frac{21}{84} =$

9) $\frac{36}{400} =$

10) $\frac{3}{11} =$

11) $\frac{24}{48} =$

12) $\frac{18}{48} =$

13) $\frac{6}{20} =$

14) $\frac{9}{125} =$

15) $\frac{36}{120} =$

16) $\frac{15}{20} =$

17) $\frac{73}{100} =$

18) $\frac{9}{45} =$

19) $\frac{9}{10} =$

20) $\frac{6}{60} =$

21) $\frac{7}{42} =$

22) $\frac{11}{88} =$

Answer key Chapter 7

Graph Decimals

1) a. 0.432 b. 0.44 c. 0.444 d. 0.452
2) a. 0.577 b. 0.581 c. 0.585 d. 0592
3) a. 0.714 b. 0.718 c. 0.722 d. 0.731
4) a. 0.693 b. 0.70 c. 0.704 d. 0.713

Round Decimals

1) 0.9
2) 4.2
3) 8.8
4) 0.6
5) 8.0
6) 0.05
7) 9.4
8) 25.33
9) 4.63
10) 10.5
11) 3.3
12) 4.2
13) 6.2
14) 9.16
15) 36.0
16) 48.0
17) 630.0
18) 577.8
19) 24.58
20) 84.0
21) 5.41
22) 86.3
23) 354.3
24) 0.93
25) 0.007
26) 8.04
27) 41.66
28) 19.08

Decimals Addition

1) 63.35
2) 0.85
3) 33.99
4) 64.41
5) 31.11
6) 10.817
7) 97
8) 86.6
9) 54.2
10) 24.33
11) 50.481
12) 41.48
13) 51.82
14) 130.12

Decimals Subtraction

1) 5.73
2) 30.07
3) 0.39
4) 16.554
5) 0.729
6) 20.73
7) 62.88
8) 26.08
9) 17.71
10) 8.099
11) 19.09
12) 11.01
13) 26.082
14) 136.64

Decimals Multiplication

1) 4.56
2) 18.72
3) 8.2008

4) 459
5) 4,980
6) 174.552
7) 250.002
8) 13.8915
9) 1.2645
10) 206.1612
11) 180.544
12) 0.3824
13) 0.825
14) 92.862

Decimal Division

1) 0.007
2) 0.0003
3) 0.28
4) 0.0008
5) 0.2
6) 0.833….
7) 0.2
8) 0.05
9) 0.0009
10) 1.111…
11) 100
12) 2
13) 1.25
14) 1,000
15) 0.049
16) 0.073
17) 0.85
18) 3.68181…
19) 11.1818…
20) 10
21) 1.19428…
22) 0.3
23) 0.125
24) 0.00167
25) 10.467
26) 0.002

Comparing Decimals

1) <
2) >
3) =
4) <
5) >
6) <
7) >
8) <
9) >
10) >
11) <
12) <
13) <
14) >
15) >
16) =
17) <
18) <
19) <
20) >
21) <
22) >
23) >
24) >
25) >
26) =
27) >
28) <

Convert Fraction to Decimal

1) 0.6
2) 0.58
3) 0.8
4) 0.125
5) 0.08
6) 0.07
7) 0.0625
8) 0.25
9) 0.09
10) 0.27
11) 0.5
12) 0.375
13) 0.3
14) 0.072
15) 0.3

16) 0.75
17) 0.73
18) 0.2

19) 0.9
20) 0.1
21) 0.167

22) 0.125

Chapter 8: Measurement

Reference Measurement

LENGTH	
Customary	Metric
1 mile (mi) = 1,760 yards (yd)	1 kilometer (km) = 1,000 meters (m)
1 yard (yd) = 3 feet (ft)	1 meter (m) = 100 centimeters (cm)
1 foot (ft) = 12 inches (in.)	1 centimeter(cm) = 10 millimeters(mm)
VOLUME AND CAPACITY	
Customary	Metric
1 gallon (gal) = 4 quarts (qt)	1 liter (L) = 1,000 milliliters (mL)
1 quart (qt) = 2 pints (pt.)	
1 pint (pt.) = 2 cups (c)	
1 cup (c) = 8 fluid ounces (Fl oz)	
WEIGHT AND MASS	
Customary	Metric
1 ton (T) = 2,000 pounds (lb.)	1 kilogram (kg) = 1,000 grams (g)
1 pound (lb.) = 16 ounces (oz)	1 gram (g) = 1,000 milligrams (mg)
Time	
1 year = 12 months	
1 year = 52 weeks	
1 week = 7 days	
1 day = 24 hours	
1 hour = 60 minutes	
1 minute = 60 seconds	

Metric Length Measurement

Convert to the units.

1) 2,000 mm = _____ cm

2) 5 m = _____ mm

3) 7 m = _____ cm

4) 9 km = _____ m

5) 5,000 mm = _____ m

6) 2,800 cm = _____ m

7) 13 m = _____ cm

8) 4,000 mm = _____ cm

9) 20,000 mm = _____ m

10) 7 km = _____ mm

11) 6 km = _____ m

12) 3 m = _____ cm

13) 17,000 m = _____ km

14) 500,000 m = _____ km

Customary Length Measurement

Convert to the units.

1) 15 ft = _____ in

2) 8 ft = _____ in

3) 7 yd = _____ ft

4) 9 yd = _____ ft

5) 3 yd = _____ in

6) 3 mi = _____ in

7) 7,200 in = _____ yd

8) 252 in = _____ yd

9) 8,800 yd = _____ mi

10) 12 yd = _____ in

11) 4 mi = _____ yd

12) 47,520 ft = _____ mi

13) 60 in = _____ ft

14) 25 yd = _____ ft

15) 36 in = _____ ft

16) 2 mi = _____ ft

Metric Capacity Measurement

Convert the following measurements.

1) 50 l = _____ ml

2) 4 l = _____ ml

3) 13 l = _____ ml

4) 8 l = _____ ml

5) 19 l = _____ ml

6) 2 l = _____ ml

7) 70,000 ml = _____ l

8) 8,000 ml = _____ l

9) 37,000 ml = _____ l

10) 200,000 ml = _____ l

11) 6,000,000 ml = _____ l

12) 40,000 ml = _____ l

Customary Capacity Measurement

Convert the following measurements.

1) 2 gal = _____ qt.

2) 11 gal = _____ pt.

3) 3 gal = _____ c.

4) 14 pt. = _____ c

5) 43 c = _____ fl oz

6) 16 qt = _____ pt.

7) 8 qt = _____ c

8) 29 pt. = _____ c

9) 6,720 c = _____ gal

10) 144 pt. = _____ gal

11) 72 qt = _____ gal

12) 92 pt. = _____ qt

13) 4,600 c = _____ qt

14) 146 c = _____ pt.

15) 108 qt = _____ gal

16) 1,848 pt. = _____ qt

17) 31 gal = _____ pt.

18) 6 qt = _____ c

19) 640 c = _____ gal

20) 104 fl oz = _____ c

Metric Weight and Mass Measurement

Convert.

1) 7 kg = _____ g

2) 3 kg = _____ g

3) 13 kg = _____ g

4) 21 kg = _____ g

5) 9 kg = _____ g

6) 121 kg = _____ g

7) 249 kg = _____ g

8) 4,000 g = _____ kg

9) 6,000 g = _____ kg

10) 17,000 g = _____ kg

11) 129,000 g = _____ kg

12) 220,000 g = _____ kg

13) 9,000,000 g = _____ kg

14) 11,000,000 g = _____ kg

Customary Weight and Mass Measurement

Convert.

1) 16,000 lb. = _____ T

2) 20,000 lb. = _____ T

3) 170,000 lb. = _____ T

4) 44,000 lb. = _____ T

5) 7 lb. = _____ oz

6) 4 lb. = _____ oz

7) 10 lb. = _____ oz

8) 24 T = _____ lb.

9) 3 T = _____ lb.

10) 9 T = _____ lb.

11) 112 T = _____ lb.

12) 2 T = _____ oz

13) 5 T = _____ oz

14) 224 oz = _____ lb.

Time

Convert to the units.

1) 16 hr. = _____ min

2) 9 year = _____ week

3) 2 hr. = _____ sec

4) 12 min = _____ sec

5) 600 min = _____ hr

6) 730 day = _____ year

7) 3 year = _____ hr.

8) 27 day = _____ hr

9) 4 day = _____ min

10) 540 min = _____ hr

11) 16 year = _____ month

12) 19,200 sec = _____ min

13) 288 hr = _____ day

14) 23 weeks = _____ day

How much time has passed?

15) From 4:15 A.M. to 7:25 A.M.: ____ hours and ____ minutes.

16) From 3:40 A.M. to 8:25 A.M.: ____ hours and ____ minutes.

17) It's 8:50 P.M. What time was 2 hours ago? _____ O'clock

18) 3:20 A.M to 6:40 AM: _____ hours and _____ minutes.

19) 3:30 A.M to 6:05 AM: _____ hours and _____ minutes.

20) 7:10 A.M. to 8:15 AM. = _____ hour(s) and _____ minutes.

21) 11:55 A.M. to 4:25 PM. = _____ hour(s) and _____ minutes

22) 7:18 A.M. to 7:52 A.M. = _____ minutes

23) 9:13 A.M. to 9:50 A.M. = _____ minutes

Answers of Worksheets – Chapter 8

Metric length

1) 200 cm
2) 5,000 mm
3) 700 cm
4) 9,000 m
5) 5 m
6) 28 m
7) 1,300 cm
8) 400 cm
9) 20 m
10) 7,000,000 mm
11) 6,000 m
12) 300 cm
13) 17 km
14) 500 km

Customary Length

1) 180
2) 96
3) 21
4) 27
5) 108
6) 190,080
7) 200
8) 7
9) 5
10) 432
11) 7,040
12) 9
13) 5
14) 75
15) 3
16) 10,560

Metric Capacity

1) 50,000 ml
2) 4,000 ml
3) 13,000 ml
4) 8,000 ml
5) 19,000 ml
6) 2,000 ml
7) 70 L
8) 8 L
9) 37 L
10) 200 L
11) 6,000 L
12) 40 L

Customary Capacity

1) 8 qt
2) 88 pt.
3) 48 c
4) 28 c
5) 344 fl oz
6) 12 pt.
7) 32 c
8) 58 c
9) 420 gal
10) 18 gal
11) 18 gal
12) 46 qt
13) 1,150 qt
14) 73 pt.
15) 27 gal
16) 927 qt
17) 248 pt.
18) 24 c
19) 40 gal
20) 13 c

Metric Weight and Mass

1) 7,000 g
2) 3,000 g
3) 13,000 g
4) 21,000 g
5) 9,000 g
6) 121,000 g
7) 249,000 g
8) 4 kg
9) 6 kg
10) 17 kg
11) 129 kg
12) 220 kg
13) 9,000 kg
14) 11,000 kg

Customary Weight and Mass

1) 8 T
2) 10 T
3) 85 T
4) 22 T
5) 112 oz
6) 64 oz
7) 160 oz
8) 48,000 lb.
9) 6,000 lb.
10) 18,000 lb.
11) 224,000 lb.
12) 64,000 oz
13) 160,000 oz
14) 14 lb

Time

1) 960 min
2) 468 weeks
3) 7,200 sec
4) 720 sec
5) 10 hr
6) 2 year
7) 26,280 hr
8) 648 hr
9) 5,760 min
10) 9 hr
11) 192 months
12) 320 min
13) 12 days
14) 161 days
15) 3:10
16) 4:45
17) 6:50 P.M.
18) 3:20
19) 2:35
20) 1:05
21) 4:30
22) 34 minutes
23) 37 minutes

Chapter 9: Symmetry and Transformations

Line Segments

Write each as a line, ray, or line segment.

1)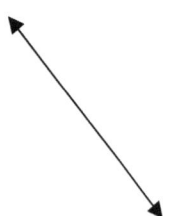

2)

3)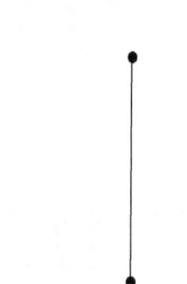

4)

5)

6)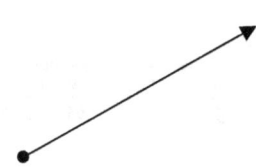

7)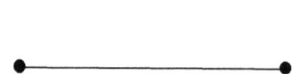

8)

Parallel, Perpendicular and Intersecting Lines

State whether the given pair of lines are parallel, perpendicular, or intersecting.

1)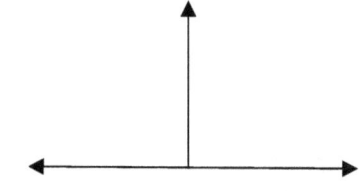

2)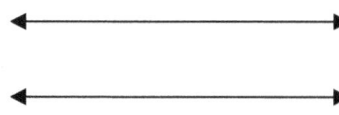

3)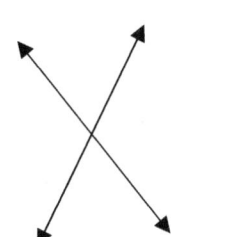

4)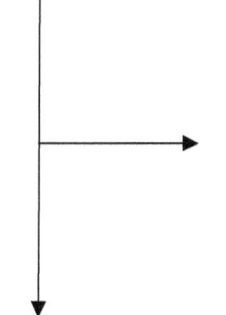

5)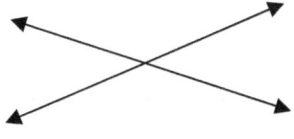

6)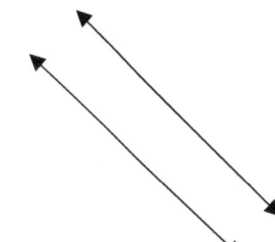

7)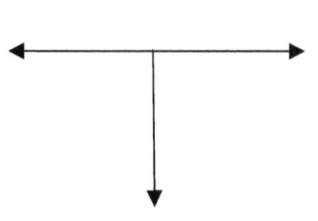

8)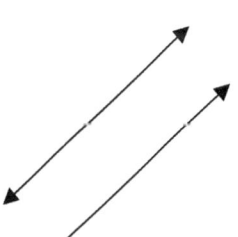

Identify Lines of Symmetry

Tell whether the line on each shape a line of symmetry is.

1)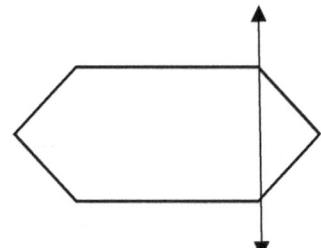

2)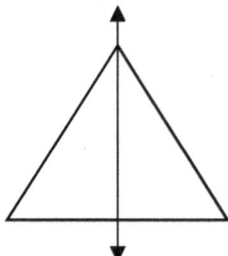

3)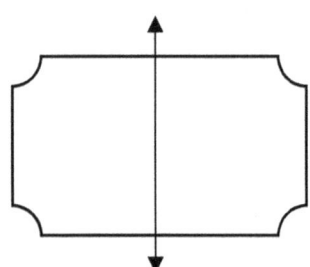

4)

5)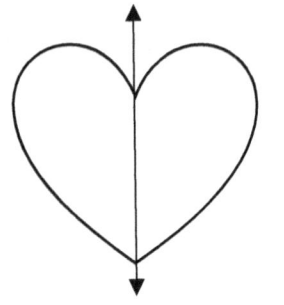

6)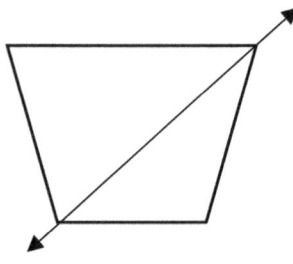

7)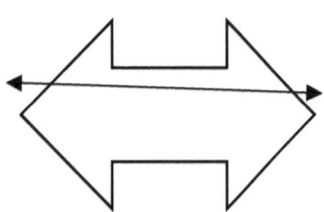

8)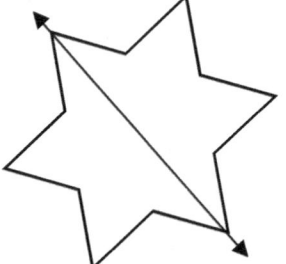

Lines of Symmetry

Draw lines of symmetry on each shape. Count and write the lines of symmetry you see.

1)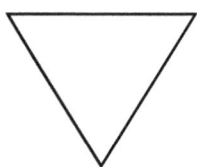

2)

3)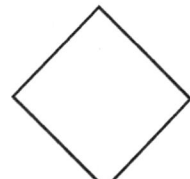

4)

5)

6)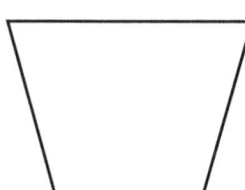

7)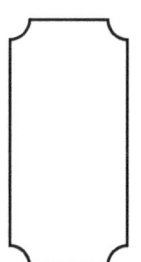

8)

Identify Three–Dimensional Figures

Write the name of each shape.

1)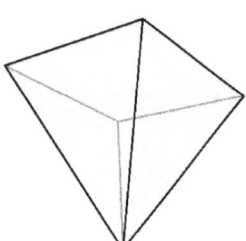

2)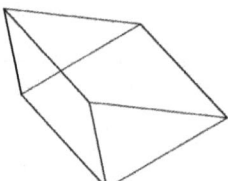

3)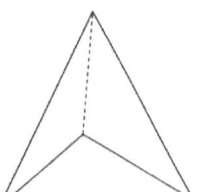

4)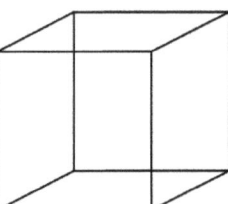

5)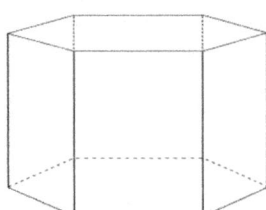

6)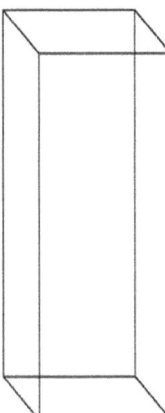

7)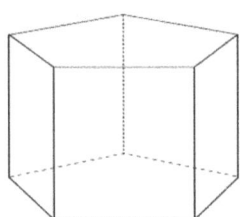

Vertices, Edges, and Faces

Complete the chart below.

	Shape	Number of edges	Number of faces	Number of vertices
1)	hexagonal prism	_____	_____	_____
2)	rectangular prism	_____	_____	_____
3)	tetrahedron	_____	_____	_____
4)	square pyramid	_____	_____	_____
5)	cube	_____	_____	_____
6)	pentagonal prism	_____	_____	_____

Identify Faces of Three-Dimensional Figures

Write the number of faces.

1)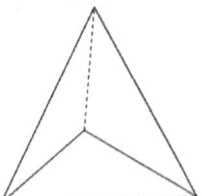

2)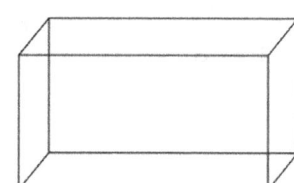

3)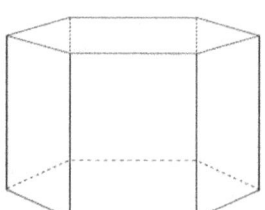

4)

5)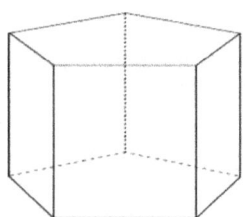

6)

7)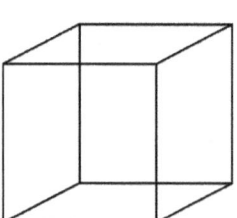

8)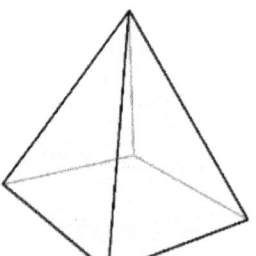

Answers of Worksheets – Chapter 9

Line Segments

1) Line
2) Line segment
3) Line segment
4) Ray
5) Line
6) Ray
7) Line segment
8) Ray

Parallel, Perpendicular and Intersecting Lines

1) Perpendicular
2) Parallel
3) Intersection
4) Perpendicular
5) Intersection
6) Parallel
7) Perpendicular
8) Parallel

Identify lines of symmetry

1) No
2) yes
3) yes
4) No
5) yes
6) No
7) No
8) yes

lines of symmetry

1)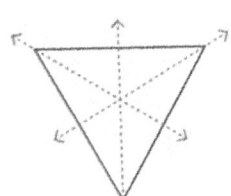

2)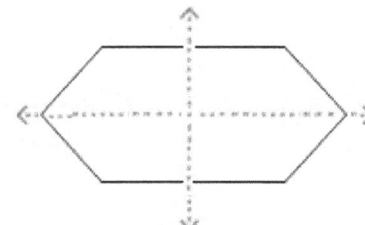

3)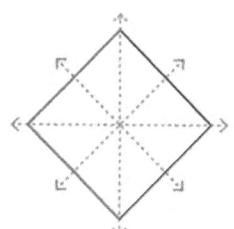

4)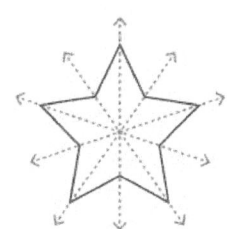

5)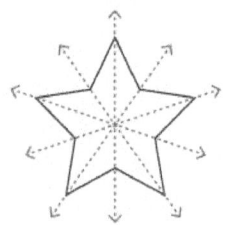
6)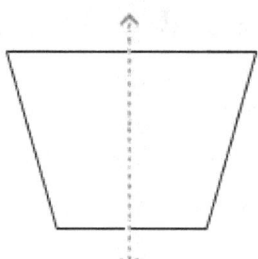

7)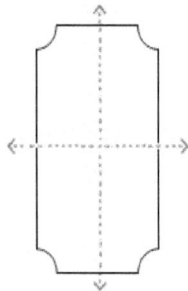
8)

Identify Three–Dimensional Figures

1) Square pyramid
2) Triangular prism
3) Triangular pyramid
4) Cube

5) Hexagonal prism
6) Rectangular prism
7) Pentagonal prism

Vertices, Edges, and Faces

Shape	Number of edges	Number of faces	Number of vertices
1)	18	8	12
2)	12	6	8
3)	6	4	4

PSSA Math Practice Grade 4

4) 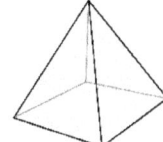   8   5   5

5) 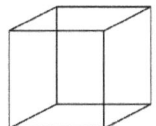   12   6   8

6) 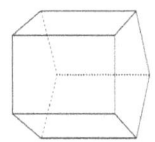   15   7   10

Identify Faces of Three–Dimensional Figures

1) 4   4) 2   7) 6
2) 6   5) 7   8) 5
3) 8   6) 5

Chapter 10:

Geometry

Identifying Angles

Write the name of the angles(Acute, Right, Obtuse, and Straight Angles) .

1)

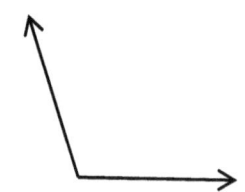

2)

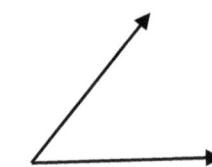

3)

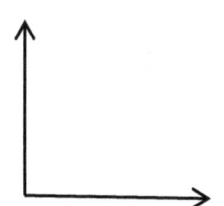

4)

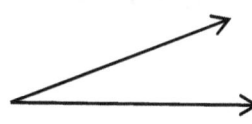

5)

6)

7)

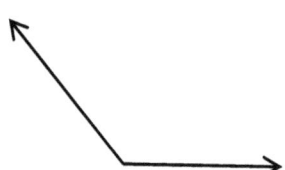

8)

Polygon Names

Write name of polygons.

1)
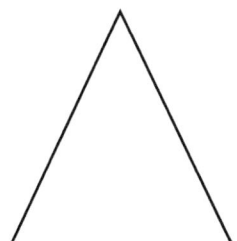

2)

3)
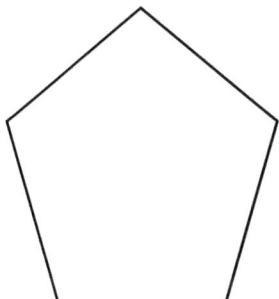

4)
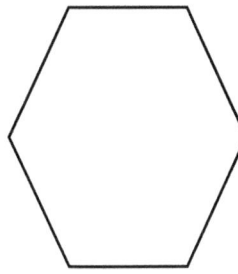

5)
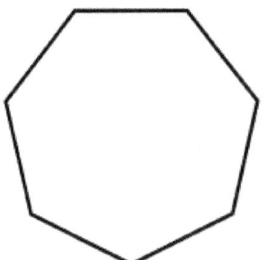

6)
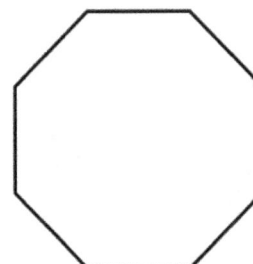

7)
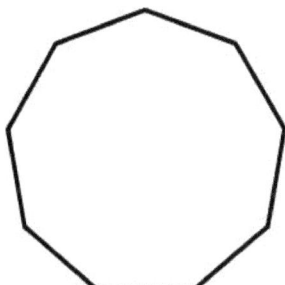

8)
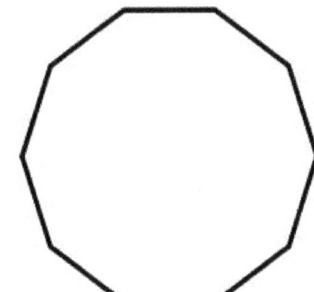

Triangles

Classify the triangles by their sides and angles.

1)

2)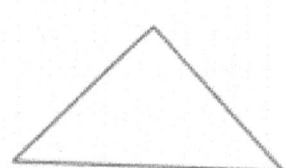

3)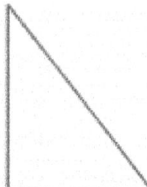

4)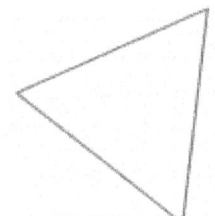

5)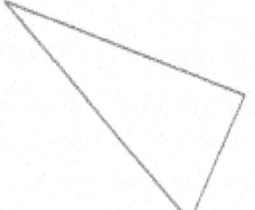

6)

Find the measure of the unknown angle in each triangle.

7)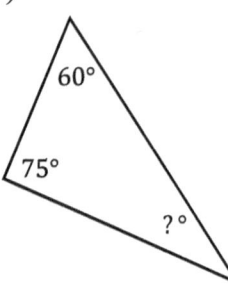

8)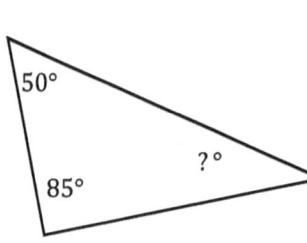

9)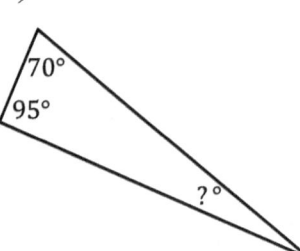

10)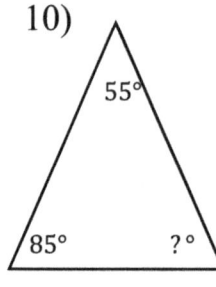

11)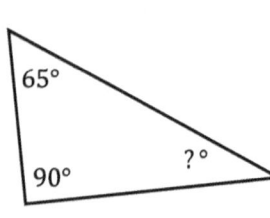

12)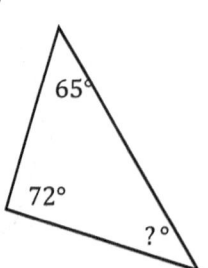

13)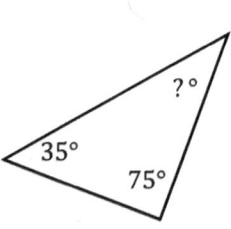

14)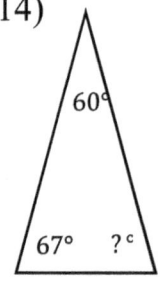

Quadrilaterals and Rectangles

Write the name of quadrilaterals.

1)

2)

3)

4)

5)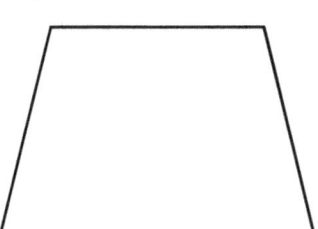

6)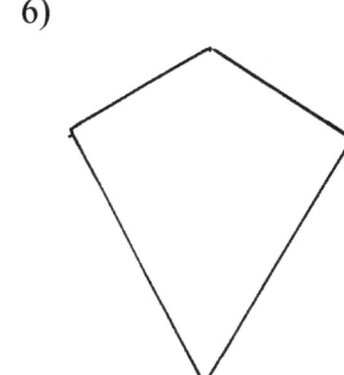

Solve.

7) A rectangle has _____ sides and _____ angles.

8) Draw a rectangle that is 6 centimeters long and 5 centimeters wide. What is the perimeter?

9) Draw a rectangle 5 cm long and 3 cm wide.

10) Draw a rectangle whose length is 6cm and whose width is 4 cm. What is the perimeter of the rectangle?

11) What is the perimeter of the rectangle?

Area and Perimeter of Square

Find the perimeter and area of each squares.

1)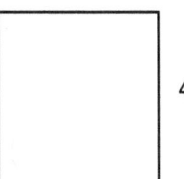

Perimeter: .... :

Area: ....

2)

Perimeter: .... :

Area: .... :

3)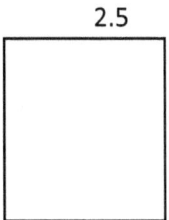

Perimeter: .... :

Area: .... :

4)

Perimeter: .... :

Area: .... :

5)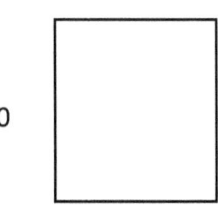

Perimeter: .... :

Area: .... :

6)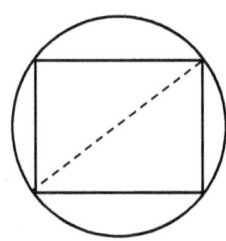

Perimeter of Square: .... :

Area of Square: .... :

Area and Perimeter of Rectangle

Find the perimeter and area of each rectangle.

1)

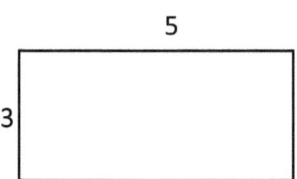

Perimeter: ....

Area: ....

2)

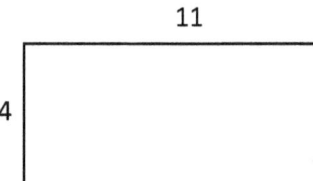

Perimeter: ....

Area: ....

3)

[Rectangle with top side 10 and left side 1.2]

Perimeter: ....

Area: ....

4)

[Rectangle with top side $9\frac{1}{2}$ and left side 2]

Perimeter: ....

Area: ....

5)

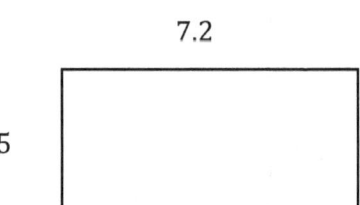

Perimeter: ....

Area: ....

6)

[Rectangle with top side 8.4 and left side 3.6]

Perimeter: ....

Area: ....

Area and Perimeter of Triangle

Find the perimeter and area of each triangle.

1)

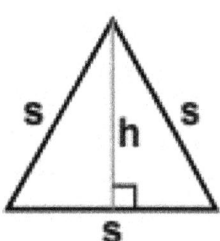

Perimeter: ....
Area: ....

2)

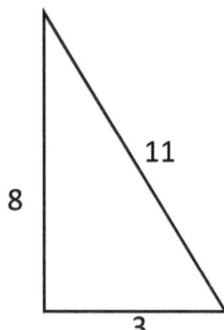

Perimeter: ....
Area: ....

3)

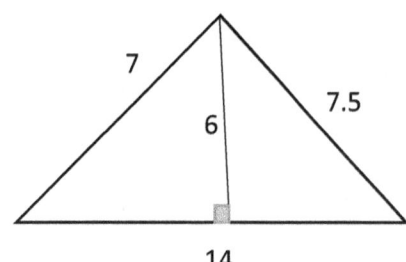

Perimeter: ....
Area: ....

4)

$s=10$

$h=6.4$

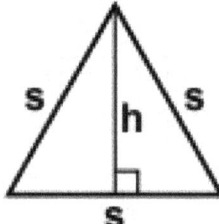

Perimeter: ....
Area: ....

5)

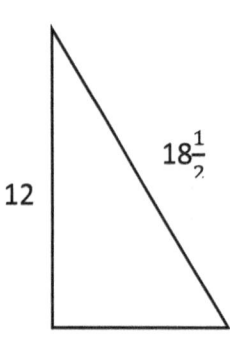

Perimeter: ....
Area: ....

6)

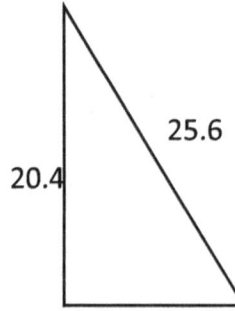

Perimeter: ....
Area: ....

Perimeter of Polygon

Find the perimeter of each polygon.

1)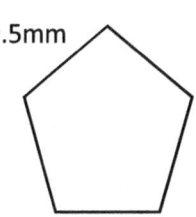

Perimeter: _____.

2)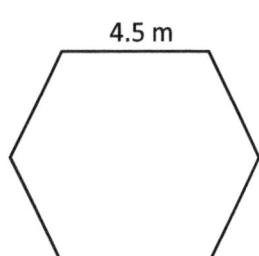

Perimeter: _____.

3)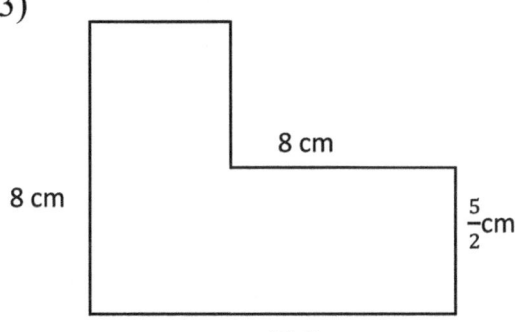

Perimeter: _____.

4)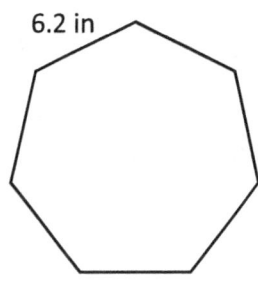

Perimeter: _____.

5)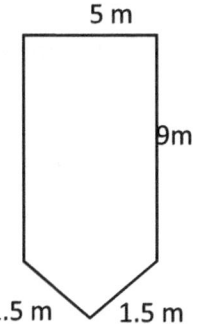

Perimeter: _____.

6)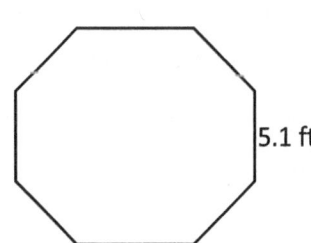

Perimeter: _____.

Answer key Chapter 10

Identifying Angles

1) Obtuse
2) Acute
3) Right
4) Acute
5) Straight
6) Obtuse
7) Obtuse
8) Acute

Polygon Names

1) Triangle
2) Quadrilateral
3) Pentagon
4) Hexagon
5) Heptagon
6) Octagon
7) Nonagon
8) Decagon

Triangles

1) Scalene, obtuse
2) Isosceles, right
3) Scalene, right
4) Equilateral, acute
5) Scalene, acute
6) Scalene, acute
7) 45°
8) 45°
9) 15°
10) 40°
11) 25°
12) 43°
13) 70°
14) 53°

Quadrilaterals and Rectangles

1) Square
2) Rectangle
3) Parallelogram
4) Rhombus
5) Trapezoid
6) Kike
7) 4 - 4
8) 22
9) Use a rule to draw the rectangle
10) 20
11) 26

Area and Perimeter of Square

1. Perimeter: 16,   Area:16
2. Perimeter: 6,   Area:2.25
3. Perimeter: 10, Area:6.25
4. Perimeter: 32,   Area:64
5. Perimeter: 40,   Area:100
6. Perimeter: $4\sqrt{3}$, Area:3

Area and Perimeter of Rectangle

1- Perimeter: 16,   Area:15
2- Perimeter: 30,   Area:44
3- Perimeter: 22.4, Area:12
4- Perimeter: 23, Area:19
5- Perimeter: 24.4, Area:36
6- Perimeter:24, Area:30.24

Area and Perimeter of Triangle

1- Perimeter: 3s, Area:$\frac{1}{2}sh$
2- Perimeter: 22,   Area:12
3- Perimeter: 28.5, Area:42
4- Perimeter: 30,   Area:32
5- Perimeter: 38.5, Area:48
6- Perimeter: 60, Area:142.8

Perimeter of Polygon

1) 47.5 mm
2) 27 m
3) 41 cm
4) 43.4 in
5) 26 m
6) 40.8 ft

Chapter 11: Data and Graphs

Tally and Pictographs

Using the key, draw the pictograph to show the information.

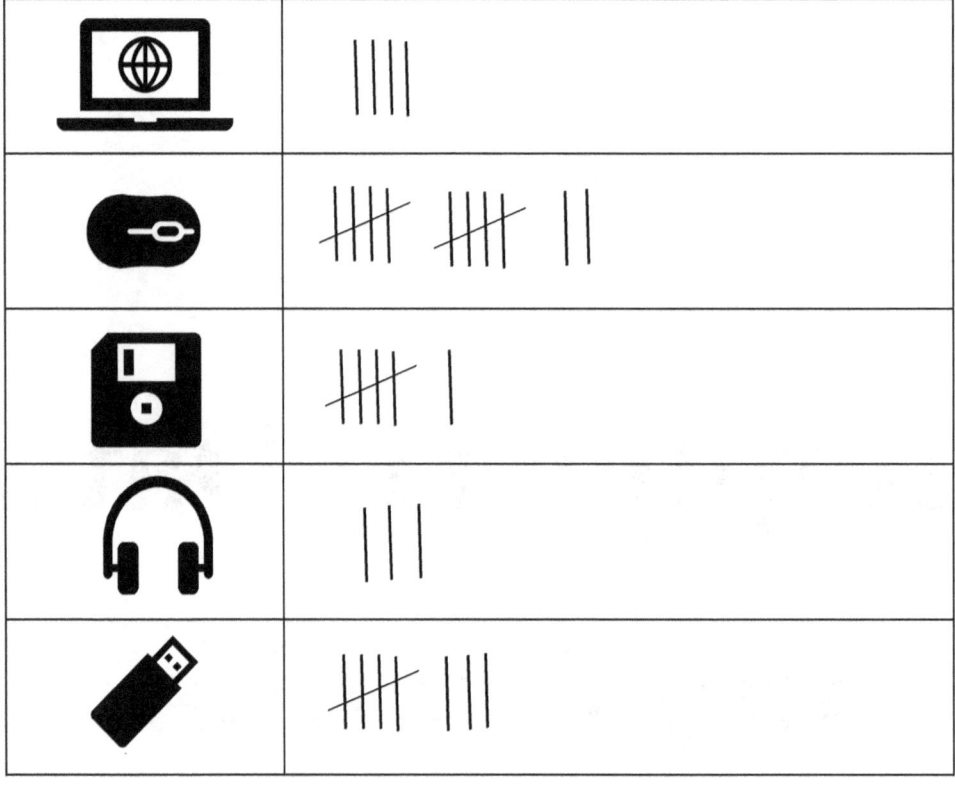

Key: ✳ = 2 Hardware

Stem–And–Leaf Plot

Make stem-and-leaf plots for the given data.

1) 15, 16, 38, 31, 12, 54, 18, 37, 39, 34, 19, 32, 55

2) 72, 74, 17, 41, 72, 14, 46, 78, 48, 44, 49, 42

3) 125, 108, 65, 65, 105, 127, 62, 126, 68, 124, 66, 109

4) 61, 45, 66, 60, 99, 63, 90, 97, 68, 63, 49, 42

5) 55, 58, 105, 56, 15, 108, 102

6) 123, 57, 77, 55, 120, 127, 73, 124, 58, 123, 79, 71

Dot plots

The ages of students in a Math class are given below.

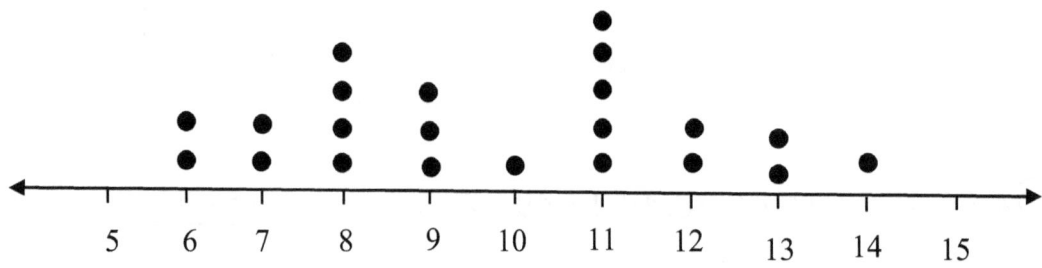

1) What is the total number of students in math class?

2) How many students are at least 12 years old?

3) Which age(s) has the most students?

4) Which age(s) has the fewest student?

5) Determine the median of the data.

6) Determine the range of the data.

7) Determine the mode of the data.

Coordinate Plane

Plot each point on the coordinate grid.

1) A (2, 7)

2) B (6, 3)

3) C (0, 7)

4) D (3, 0)

5) E (1, 4)

6) F (3, 9)

7) G (5, 1)

8) H (7, 7)

9) I (9, 6)

10) J (6, 1)

11) K (2, 4)

12) L (3, 8)

Bar Graph

Each student in class selected two games that they would like to play. Graph the given information as a bar graph and answer the questions below:

Game	Votes
Football	13
Volleyball	10
Basketball	18
Baseball	17
Tennis	13

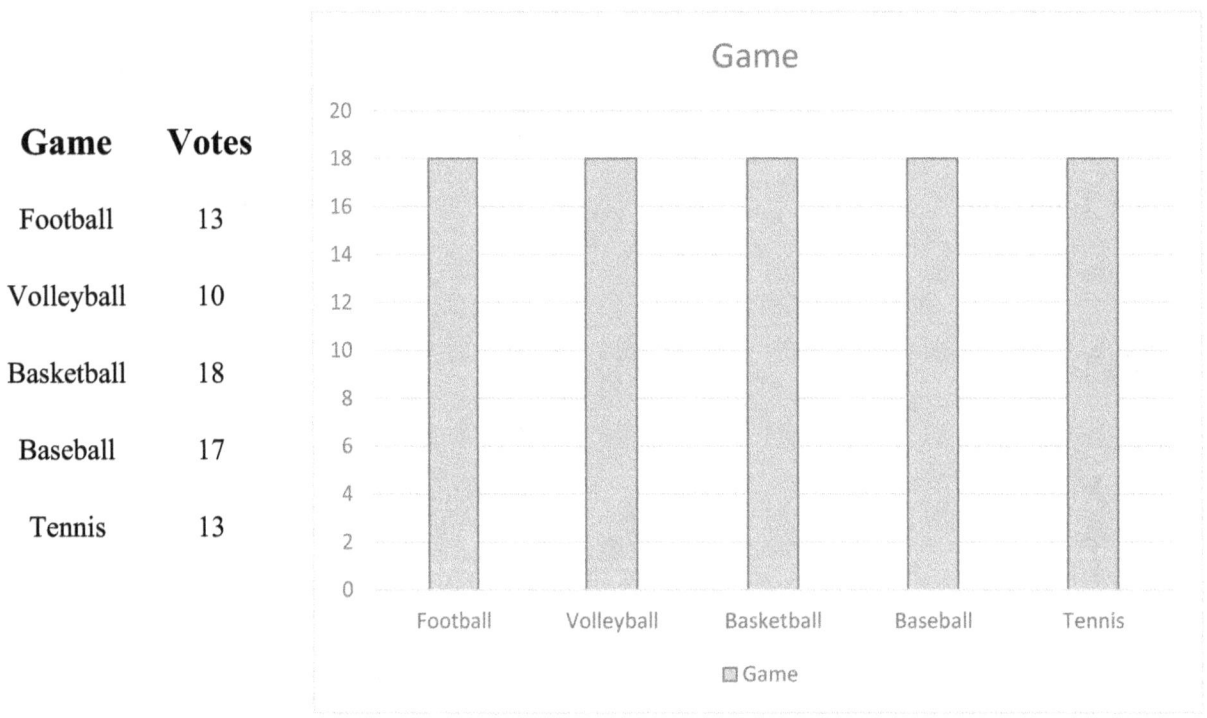

1) Which was the most popular game to play?

2) How many more students like Basketball than Volleyball?

3) Which two game got the same number of votes?

4) How many Volleyball and Football did student vote in all?

5) Did more student like football or Volleyball?

6) Which game did the fewest student like?

Line Graphs

Amelia works in a doll store. She records the number of dolls sold in five days on a line graph. Use the graph to answer the questions.

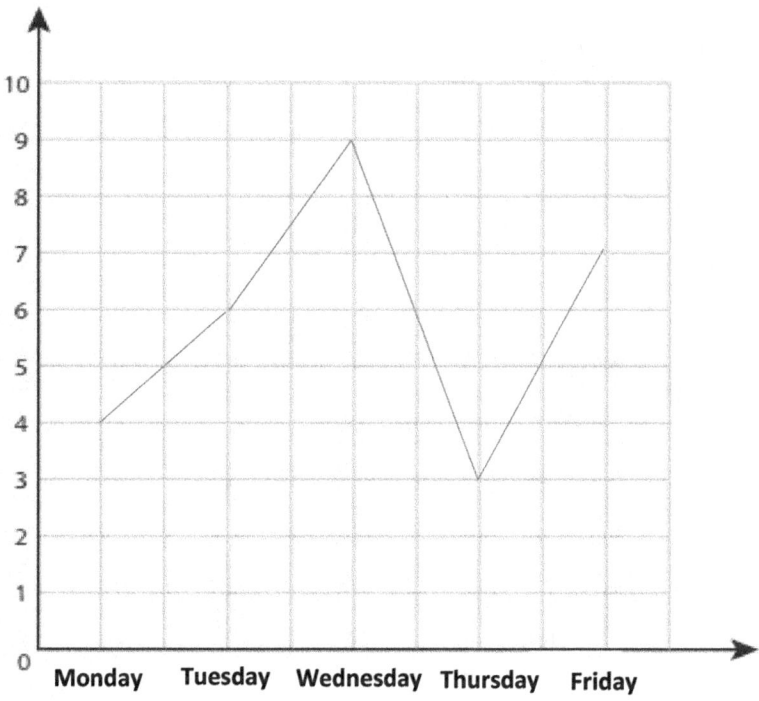

1) How many dolls were sold on Tuesday?

2) Which day had the minimum sales of dolls?

3) Which day had the maximum number of dolls sold?

4) How many dolls were sold in 5 days?

Answer key Chapter 11

Tally and Pictographs

💻	※ ※
🖱	※ ※ ※ ※ ※ ※
💾	※ ※ ※
🎧	※ ⌄
🔌	※ ※ ※ ※

Stem–And–Leaf Plot

1)

Stem	leaf
1	2 5 6 8 9
3	1 2 4 7 8 9
5	4 5

2)

Stem	leaf
1	4 7
4	1 2 4 6 8 9
7	2 2 4 8

3)

Stem	leaf
6	2 5 5 6 8
10	5 8 9
12	4 5 6 7

4)

Stem	leaf
4	2 9 5
6	0 1 3 3 6 8
9	0 7 9

5)

Stem	leaf
1	5
5	5 6 8
10	2 5 8

6)

Stem	leaf
5	5 7 8
7	1 3 7 9
12	0 3 3 4 7

Dot plots

1) 22
2) 5
3) 11
4) 10 and 14
5) 2
6) 4
7) 2

PSSA Math Practice Grade 4

Graph Points on a Coordinate Plane

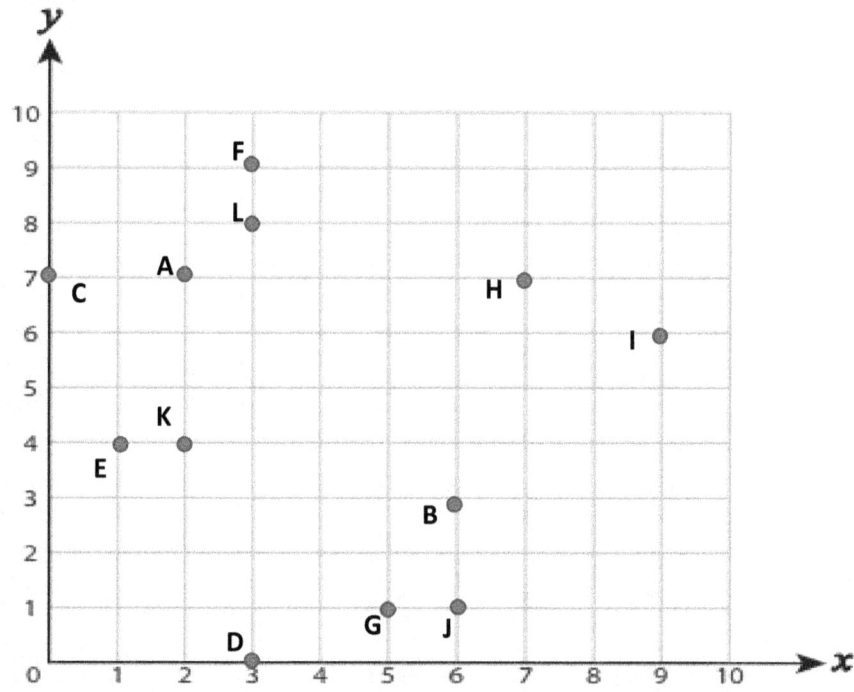

Bar Graph

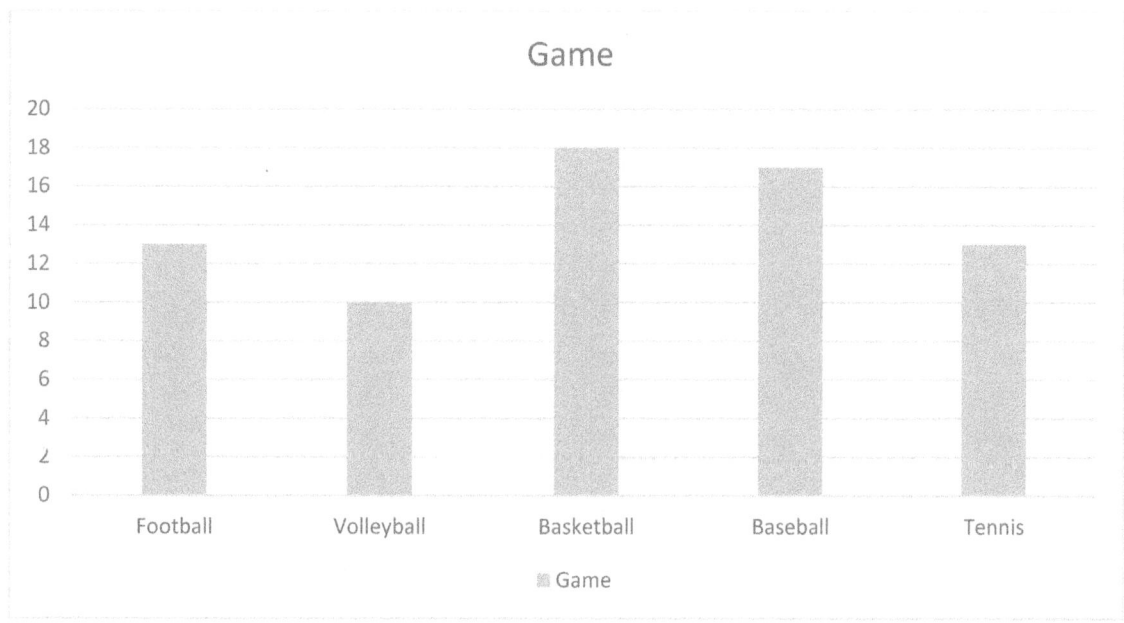

1) Basketball   3) Football and Tennis   5) Football
2) 8 students   4) 23   6) Volleyball

Line Graphs

1) 6   2) Thursday   3) Wednesday   4) 29

PSSA Test Review

PSSA GRADE 4 MAHEMATICS REFRENCE MATERIALS

LENGTH

Customary	Metric
1 mile (mi) = 1,760 yards (yd)	1 kilometer (km) = 1,000 meters (m)
1 yard (yd) = 3 feet (ft)	1 meter (m) = 100 centimeters (cm)
1 foot (ft) = 12 inches (in.)	1 centimeter (cm) = 10 millimeters (mm)

VOLUME AND CAPACITY

Customary	Metric
1 gallon (gal) = 4 quarts (qt)	1 liter (L) = 1,000 milliliters (mL)
1 quart (qt) = 2 pints (pt.)	
1 pint (pt.) = 2 cups (c)	
1 cup (c) = 8 fluid ounces (Fl oz)	

WEIGHT AND MASS

Customary	Metric
1 ton (T) = 2,000 pounds (lb.)	1 kilogram (kg) = 1,000 grams (g)
1 pound (lb.) = 16 ounces (oz)	1 gram (g) = 1,000 milligrams (mg)

Time

1 year = 12 months	1 day = 24 hours
1 year = 52 weeks	1 hour = 60 minutes
1 week = 7 days	1 minute = 60 seconds

Perimeter

Square $\qquad$ $P = 4S$

Rectangle $\qquad$ $P = L + W + L + W$ or $P = 2L + 2W$

Area

Square $\qquad$ $A = S \times S$

Rectangle $\qquad$ $A = L \times W$

WWW.MathNotion.com

The Pennsylvania System of School Assessment

PSSA Practice Test 1

Mathematics

GRADE 4

❖ 20 Questions

❖ Calculators are not permitted for this practice test

Pennsylvania Department of Education Bureau of Curriculum, Assessment, and Instruction— *Month Year*

1) Which decimal is equivalent to $\frac{8}{1,000}$?

   A. 0.08

   B. 0.008

   C. 0.8

   D. 8.00

2) In which drawing does line A appear to be perpendicular to line B?

   A.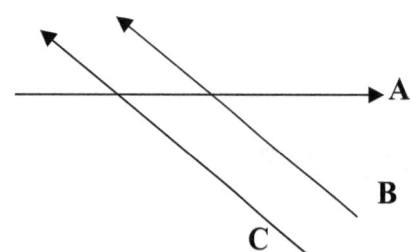

   B.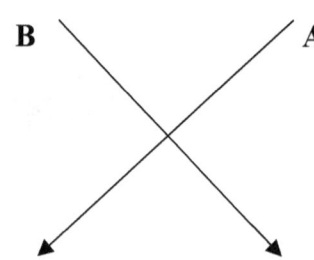

   C.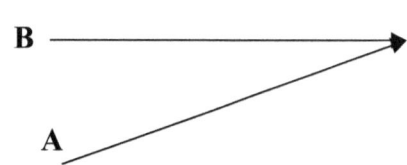

   D.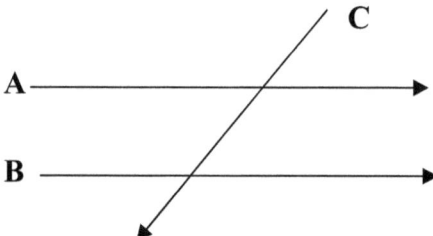

3) Which digit can replace the box below to make the comparing true?

$$7,619 < 7,\boxed{}91 < 7,801$$

   A. 8

   B. 7

   C. 0

   D. 9

PSSA Math Practice Grade 4

4) Nancy makes mango smoothies for a party. She combines 5 gallons 3 quarts of milk with 3 gallons 3 quarts of mango juice. Her sister adds 2 gallons 2 quarts of the smoothies. How much is all the mango smoothies?

   A. 11 gallons 3 quart

   B. 9 gallons 4 quarts

   C. 12 gallons

   D. 10 gallons 3 quarts

5) Emma had 5 feet of yarn. She needed 3 inches to make a bracelet. How many bracelets can she make with the amount of yarn she has?

   A. 20

   B. 18

   C. 30

   D. 60

6) Sam has 1,500 dimes. Anna has $\frac{1}{15}$ number of dimes that Sam does. George has $\frac{1}{5}$ the number of dimes that Anna has. How much money does George have?

   A. $0.2

   B. $2.00

   C. $10

   D. $100

PSSA Math Practice Grade 4

7) Which expression can be used to solve the equation 2,500 ÷ 5 = __?__ .

   A. (25 ÷ 5) + (500 ÷ 5)

   B. (200 ÷ 5) − (500 ÷ 5)

   C. (2,000 ÷ 5) + (500 ÷ 5)

   D. (2,000 ÷ 5) − (500 ÷ 5)

8) A company used 42,098.07 feet of electrical wires last month. What is the expanded form of the number?

   A. (4 × 10,000) + (2 × 1,000) + (9 × 100) + (8 × 10) + (7 × 0.1)

   B. (4 × 10,000) + (2 × 1,000) + (9 × 10) + (8 × 1) + (7 × 0.1)

   C. (4 × 10,000) + (2 × 100) + (9 × 10) + (8 × 1) + (7 × 0.01)

   D. (4 × 10,000) + (2 × 1,000) + (9 × 10) + (8 × 1) + (7 × 0.01)

9) Select these polygons that have at least one set of parallel sides.

   Figure 1   Figure 2   Figure 3   Figure 4

   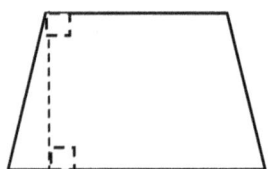
   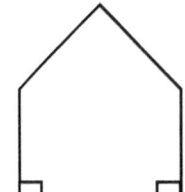

   A. Figure 1 and Figure 2

   B. Figure 3 and Figure 4

   C. Figure 1, Figure 2, and Figure 3

   D. All figures.

10) The width of a rectangular yard is 36 feet and the length is 115 feet. A tent that is 14 feet long and 10 feet wide cover part of the field. How many square feet of the field are not covered by the tent?

A. 4,120 square feet

B. 4,000 square feet

C. 3,420 square feet

D. 140 square feet

11) In which model could the shaded parts equivalent to $7 \times \frac{1}{6}$?

A.

B.

C.

D.

12) Russel has 6 acres of land and he grows vegetables on one-fifth of it. Philip has 8 acres and one-sixth of it is planted with vegetables. Who has the biggest area of vegetables?

A. Philip.

B. They have the same area.

C. Russel

D. Need more information.

13) The Ethan's family are on the road trip. They travel 85.35 miles the first day, 112.72 miles the second day, and 124.7 miles the final day. How many miles does the Ethan's family travel during the three-day trip?

A. 327.27 miles

B. 322.77 miles

C. 320.72 miles

D. 302.07 miles

14) Maria starts practicing her guitar at 8:25 A.M. she practices for 3 hours and 50 minutes and then stop 55 minutes to eat lunch. What time does Maria lunch end?

  i.   12:20 P.M.

 ii.   11:20 A.M.

iii.   12:15 P.M.

iv.   13:10 P.M.

15) What is the measure of angle DOC to the nearest degree?

A. 70°

B. 65°

C. 110°

D. 140°

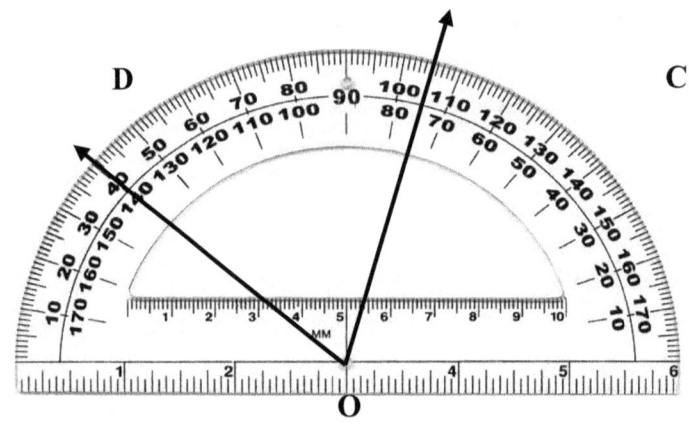

16) The data shows the 30 hours rain by each of 5 days. Which frequency table represents the number of hours rain each day?

3 hours rain Monday

5 fewer hours rain Wednesday than Tuesday.

8 hours rain Tuesday

3 times as many as Monday rain Friday.

The rest of hours rain Thursday.

A.

Hours of Rain	
Day	Number of hours
Monday	IIII
Tuesday	IIII II
Wednesday	III
Thursday	IIII
Friday	IIII I

B.

Hours of Rain	
Day	Number of hours
Monday	III
Tuesday	IIII I
Wednesday	IIII IIII
Thursday	IIII I
Friday	II I

C.

Hours of Rain	
Day	Number of hours
Monday	III
Tuesday	IIII III
Wednesday	III
Thursday	IIII II
Friday	IIII III

D.

Hours of Rain	
Day	Number of hours
Monday	III
Tuesday	IIII III
Wednesday	III
Thursday	IIII I
Friday	IIII III

17) A baker sold 9 cakes for the same price. He received a total of $360. Each cake cost the baker $22 to make. How much money did he make on each cake?

A. $40

B. $18

C. $22

D. $20

18) An equation that uses fraction model is shown. Which fraction makes the equation true?

A. $\frac{45}{90}$

B. $\frac{90}{100}$

C. $\frac{1}{10}$

D. $\frac{40}{100}$

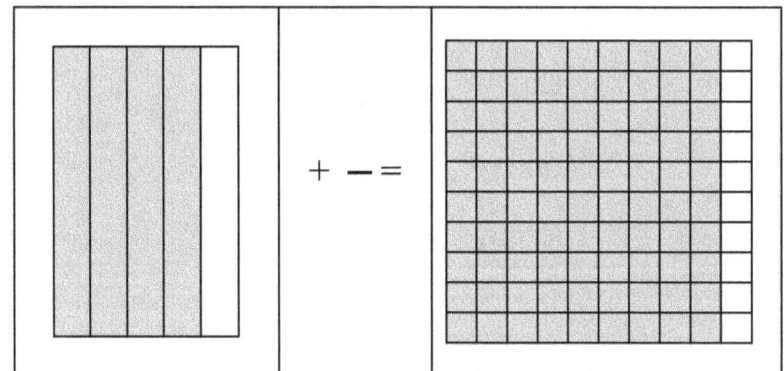

19) Ava is making a pattern for quilt. The pattern shows 40 squares. Every 5rd square is green. How many green squares are in the pattern?

A. 8

B. 10

C. 5

D. 15

20) Emily write a number.

- The number is between 60 and 70.

- It has exactly 7 factors.

- One of the factors is 8.

- What is Emily's number

A. 64

B. 62

C. 68

D. 59

The Pennsylvania System of School Assessment

PSSA Practice Test 2

Mathematics

GRADE 4

- ❖ 20 Questions
- ❖ Calculators are not permitted for this practice test

Pennsylvania Department of Education Bureau of Curriculum, Assessment, and Instruction— *Month Year*

PSSA Math Practice Grade 4

1) Grace write a number.

   - The digits in ones period are 207

   - The word form of the thousands period is two hundred four thousand.

   - The digit in thousands millions and hundred place are the same.

   What is the Grace's number?

   A. 706,240,027

   B. 706,204,702

   C. 720,207,207

   D. 206,204,207

2) What is the quotient for the expression 4,526 ÷ 8?

   A. 565

   B. 656 r 3

   C. 565 r 6

   D. 656

3) Which fraction equivalent to 7.5?

   A. $\frac{15}{2}$

   B. $\frac{15}{100}$

   C. $\frac{10}{15}$

   D. $\frac{100}{15}$

PSSA Math Practice Grade 4

4) Thomas is adding a baseboard trim around the edge of his floor. The room is 16 feet wide and 23 feet long. There is a door opening that is 4 feet wide. How many feet of trim will Thomas need?

   A. 72 feet

   B. 74 feet

   C. 437 feet

   D. 48 feet

5) Which of the following fractions is higher than $\frac{1}{12}$, but less than $\frac{1}{3}$?

   A. $\frac{1}{4}$

   B. $\frac{3}{5}$

   C. $\frac{5}{6}$

   D. $\frac{4}{5}$

6) Taye had four $7 bill, five nickels, and 5 pennies. Then he bout a math book for $26.64. How much money did Taye have after bout the book?

   A. $1.59

   B. $1.64

   C. $1.66

   D. $16.60

PSSA Math Practice Grade 4

7) William has a rope 3 feet 6 inches long. He cuts it into 6 equal pieces. How many inches long is each piece?

   A. 7

   B. 8

   C. 6

   D. 10

8) Which statements represents the number sentence $28 = 7 \times 4$?

   A. 28 is 7 more than 4

   B. 28 is 7 times as many as 4

   C. 7 is 4 times as many as 28

   D. 7 more than 4 is 28

9) A restaurant manager collected data on the lengths of time customers waited for their food. His data are shown in the dot plot. How many customers waited more than 16 minutes for their food?

   A. 10

   B. 11

   C. 12

   D. 15

10) The Brandon family are on a 10-day road trip. They travel 11 hours each day for 4 days. They travel 6 hours each day for 5 days. How many hours does the Brandon family travel during their road trip?

A. 74

B. 44

C. 30

D. 132

11) An unshaded fraction model is shown. Select the equivalent to the fraction model.

A. $\frac{1}{3}$

B. $\frac{9}{15}$

C. $\frac{5}{8}$

D. $\frac{3}{7}$

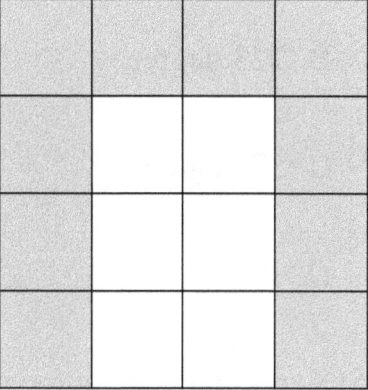

12) There are 12 books on a shelf. 4 of these books are new. What is the fraction of all books is used books?

A. $\frac{2}{9}$

B. $\frac{3}{8}$

C. $\frac{3}{5}$

D. $\frac{2}{3}$

13) Which statement correctly compares the two values?

   A. The value of 9 in 5,293 is same as the value of the 9 in 2,985

   B. The value of 9 in 5,293 is 10 the value of the 9 in 3,925

   C. The value of 9 in 3,591 is 100 times the value of the 9 in 1,953

   D. The value of 9 in 3,596 is 1000 the value of the 9 in 6,953

14) The measure of angle ABC shown below is 123 degrees. What is the measure of angle ABD?

   A. 50 degree

   B. 123 degree

   C. 57 degree

   D. 107 degree

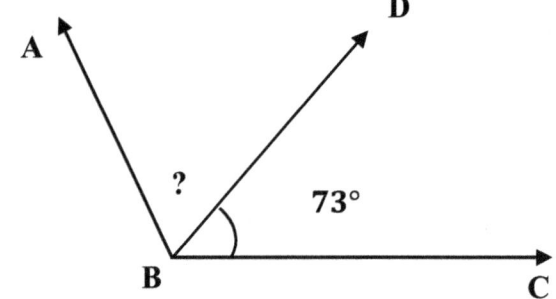

15) Kerry planted four types of beans and measured the height of the plants in centimeter after a week. Which type of bean has the shortest height?

   A. Soybean

   B. Bush Bean

   C. Lima Bean

   D. Velvet Bean

Bean	Length (cm)
Soybean	$\frac{1}{4}$
Bush Bean	$\frac{3}{5}$
Lima Bean	$\frac{5}{7}$
Velvet bean	$\frac{4}{5}$

16) Emma has 200 pony beads. Miley has 7 times as many pony beads as Emma. How many pony beads does Miley have?

A. 140

B. 1,040

C. 1,400

D. 1,004

17) Chris buys 20 candy bars at $0.80 each. If he sells them all in equal amounts for the same price to four friends, how much money will each friend give Chris?

A. $3.5

B. $4

C. $5

D. $16

18) The table shows the relation between the number of hours Ian works each week. Which could be a rule to find the hours (h) when given the week (w)?

Weeks (w)	3	4	5	6	7
Hours (h)	18	24	30	36	42

A. The output is $w - 18$

B. The output is $w \div 6$

C. The output is $w + 18$

D. The output is $w \times 6$

19) What is the measure, in degrees of an angle that is equivalent to $\frac{1}{45}$ of a circle?

A. 8

B. 135

C. 18

D. 120

20) Which number line shows the correct locations of all given values?

23.90, 22.30

A.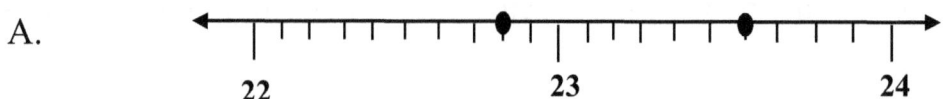

B.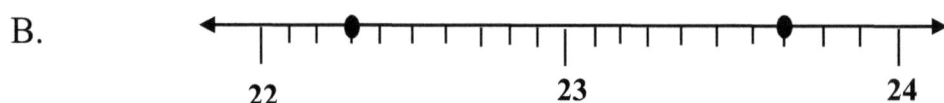

C.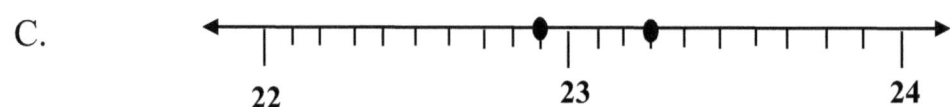

D.

Answers and Explanations

Answer Key

Now, it is time to review your results to see where you went wrong and what areas you need to improve!

PSSA Math Practice Tests

Practice Test 1

1	B	11	B
2	D	12	C
3	B	13	B
4	C	14	D
5	A	15	B
6	B	16	C
7	C	17	B
8	D	18	C
9	C	19	A
10	B	20	A

Practice Test 2

1	D	11	C
2	C	12	D
3	A	13	B
4	B	14	A
5	A	15	A
6	C	16	C
7	A	17	B
8	B	18	D
9	B	19	A
10	A	20	D

Practice Test 1

Answers and Explanations

1) Answer: B

We should have placed the 8 to the right of decimal point. The answer is 0.008

2) Answer: D

perpendicular lines are lines that intersect (cross each other) at a right angle (90° angle). Then lines A and B appear to intersect at a right angle in the option C.

3) Answer: B

when 4-digit numbers are compared start with thousands place and then hundreds place, etc. If one whole number has a higher number in the thousands place, then it is larger than a whole number with fewer thousands. If the thousands are equal compare the hundreds, then the tens etc. when compared the numbers; 7 is between 6 and 8.

4) Answer: C

Use formula: 4 quarts = 1 gallon.

5 gallons 3 quarts + 3 gallons 3 quarts = 8 gallons 6 quarts = 9 gallons 2 quart

9 gallons 2 quart + 2 gallons 2 quarts = 11 gallons 4 quarts = 12 gallons

5) Answer: A

1 foot equals to 12 inches

$5 \times 12 = 60$ inches, $60 \div 3 = 20$ barcelets

6) Answer: B

Sam: $1,500

Anna = $\frac{1}{15}$ Sam = $\frac{1}{15} \times 1,500 = \frac{1,500}{15} = 100$ dimes

George = $\frac{1}{5}$ Anna = $\frac{1}{5} \times 100 = \frac{100}{5} = 20$ dimes

Each dime is 10¢: $20 \times 10 = 200$¢ $= \$2.00$

7) Answer: C

Use fractions with same denominator: $\frac{2,500}{5} = \frac{2,000}{5} + \frac{500}{5}$

PSSA Math Practice Grade 4

8) Answer: D

In the expanded form, we break up a number according to their place value and expand it to show the value of each digit.

$42,098.07 = (4 \times 10,000) + (2 \times 1,000) + (9 \times 10) + (8 \times 1) + (7 \times 0.01)$

9) Answer: C

Parallel lines are two lines that are always the same distance apart and never touch.

10) Answer: B

Area of field: $A_1 = 115 \times 36 = 4,140$

Area of tent: $A_2 = 10 \times 14 = 140$

All area without tent: $A = A_1 - A_2 = 4,140 - 140 = 4,000$

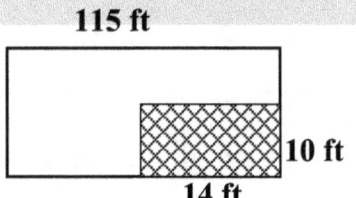

11) Answer: B

Every fraction has many equivalent fractions. $7 \times \frac{1}{6} = \frac{7}{6}$ and from the options provided:

A. $\frac{1}{5} + \frac{3}{5} = \frac{4}{5} \neq \frac{7}{6}$

B. $\frac{2}{6} + \frac{5}{6} = \frac{7}{6} = \frac{7}{6}$ is correct

C. $\frac{1}{3} + \frac{2}{3} + \frac{1}{3} = \frac{4}{3} \neq \frac{7}{6}$

D. $\frac{1}{2} + \frac{1}{2} + \frac{1}{2} = \frac{3}{2} \neq \frac{7}{6}$

12) Answer: C

Russel: $6 \times \frac{1}{5} = \frac{6}{5} = 1\frac{1}{5}$

Philip: $8 \times \frac{1}{6} = \frac{8}{6} = \frac{4}{3} = 1\frac{1}{3}$

Compare $1\frac{1}{5}$ and $1\frac{1}{3}$: $1\frac{1}{3} > 1\frac{1}{5}$, then jack has the biggest area of vegetables.

13) Answer: B

$112.72 + 124.7 + 85.35 = 322.77$

14) Answer: D

$8:25 + 3:50 = 12:15$; $12:15 + 0:55 = 13:10$ P.M

15) Answer: B

To determine the measure of the angle, you should have found the two measures on the same scale (inside or outside) through which the rays of angle pass. Then, subtract the smaller measure from the larger measure.

On the inside scale: $140° − 75° = 65°$

On the outside scale: $105° − 40° = 65°$

16) Answer: C

You need to determine the total hours of rain for each day. 3 hours rain in Monday, 8 hours rain in Tuesday, $(8 − 5) = 3$ hours rain in Wednesday, $(3 × 3) = 9$ hours rain in Friday, and $[30 − (3 + 8 + 3 + 9)] = (30 − 23) = 7$ hours rain in Thursday. Then you should have chosen the table with the same number of tally marks for each day.

17) Answer: B

$\$360 ÷ 9 = \$40$ price of each cake

Profit= Income − Cost: $\$40 − \$22 = \$18$

18) Answer: C

Express a fraction with denominator 5 as an equivalent fraction with denominator 100 and use this technique to add two fractions with respective denominators 5 and 100.

$$\frac{4}{5} + - = \frac{90}{100} \rightarrow \frac{4}{5} + - = \frac{90}{100} \rightarrow \frac{80}{100} + - = \frac{90}{100} \rightarrow - = \frac{10}{100} = \frac{1}{10}$$

19) Answer: A

$40 ÷ 5 = 8$ or

You can count them.

1.	2.	3.	4.	5.	6.	7.	8.
9.	10.	11.	12.	13.	14.	15.	16.
17.	18.	19.	20.	21.	22.	23.	24.
25.	26.	27.	28.	29.	30.	31.	32.
33.	34.	35.	36.	37.	38.	39.	40.

20) Answer: A

62 is not divisible by 8 and 59 is smaller than 60, 68 has 6 factors. Then 64 is correct.

Practice Test 2
Answers and Explanations

1) Answer: D

The word form of thousands period is: 204, and digit in thousands million and hundred place are 2.

2) Answer: C

$$\begin{array}{r} 565 \\ 8\ )\overline{4{,}526} \\ \underline{40} \\ 52 \\ \underline{48} \\ 46 \\ \underline{40} \\ 6 \end{array}$$

3) Answer: A

To convert a Decimal to a Fraction, follow these steps:

- Step 1: Write down the decimal divided by 1, like this: $\frac{7.5}{1}$
- Step 2: Multiply both top and bottom by 10 for every number after the decimal point: $\frac{75}{10}$
- Step 3: <u>Simplify</u> (or reduce) the fraction: $\frac{15}{2}$ (divided by 5).

4) Answer: B

The perimeter of a rectangle is equal to the sum of all the sides.

Perimeter is: $23 + 23 + 16 + 12 = 74$ ft

Or $P = (23 + 23) + (16 + 16) = 46 + 32 = 78, 78 - 4 = 74$

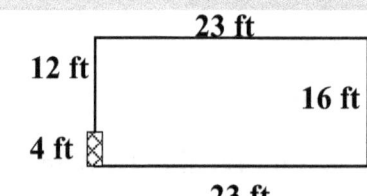

5) Answer: A

Convert these fractions to equivalent fractions with a common denominator to compare them. Use the LCD to write equivalent fractions.

PSSA Math Practice Grade 4

The LCD of (3,4,5,6,12) is 60

The equivalent fractions are: $\frac{1}{12} = \frac{5}{60}$, $\frac{1}{3} = \frac{20}{60}$

$\frac{1}{4} = \frac{15}{60}$, $\frac{3}{5} = \frac{36}{60}$, $\frac{5}{6} = \frac{50}{60}$, and $\frac{4}{5} = \frac{48}{60}$

Then, $\frac{5}{60} < \frac{?}{30} < \frac{20}{60} \rightarrow \frac{5}{60} < \frac{15}{60} < \frac{20}{60}$

6) Answer: C

First count the amount of money he had before bought the book:

$4 \times \$7 = \$28$, $5 \times \$0.05 = 0.25$,   and $5 \times \$0.01 = 0.05$

$\$28 + 0.25 + 0.05 = \$28.3$, then, subtract the price of the book:

$\$28.3 - \$26.64 = \$1.66$

7) Answer: A

Convert foot to inch: 3 feet = $3 \times 12 = 36$, $36 + 6 = 42$ inches

Then, $42 \div 6 = 7$

8) Answer: B

28 is 7 times as many as 4.

9) Answer: B

The dot plot represented the same data below:

10,10,14,14,16,18,20,20,20,22,24,24,26,26,28,28

There are 11 numbers greater than 16.

10) Answer: A

$4 \times 11 = 44$, and $6 \times 5 = 30$, then $44 + 30 = 74$ hours

11) Answer: C

The figure represents the fraction $\frac{10}{16}$, and the equivalent is: $\frac{5}{8}$ (Multiply by 2).

12) Answer: D

$12 - 4 = 8$ used book. Then, 8 books from 12 books means: $\frac{8}{12} = \frac{2}{3}$ (simplify)

13) Answer: B

The place value of 9 in 5,293 is tens

The place value of 9 in 3,925 is hundreds

To convert tens place to hundreds place we should have to multiply by 10.

14) Answer: A

$123° − 73° = 50°$

15) Answer: A

There are two main ways to compare fractions: using decimals or using the same denominator. In the decimal method, you should convert each fraction to decimals, and then compare the decimals.

$\frac{1}{4} = 0.25$, $\frac{3}{5} = 0.6$, $\frac{5}{7} = 0.714$, and $\frac{4}{5} = 0.8$

Write the decimals in order from least to greatest: $0.25, 0.6, 0.714, 0.8$

16) Answer: C

$200 \times 7 = 1,400$

17) Answer: B

$20 \times 0.8 = \$16$, and $\$16 \div 4 = \$4$

18) Answer: D

Input - the act or process of putting in.

Output - production of a certain amount.

Input is week, and output is hours, then the hours is 6 times of weeks.

19) Answer: A

There are 360 degrees in one complete circle around. $\frac{1}{45} \times 360° = \frac{360°}{45} = 8°$

20) Answer: D

First counted the number of sections on the number line between two numbers, since there are 10 section between 22, 23 and 24, each section represented one-tenths. Then for 22.30 you count 3 sections after 22 and for 23.90 you count 9 sections after 23.

"End"

www.ingramcontent.com/pod-product-compliance
Lightning Source LLC
LaVergne TN
LVHW061310060426
835507LV00019B/2096